ENDANGERED WILDLIFE IN DERBYSHIRE

Dark Red Helleborine Orchid

Endangered Wildlife in Derbyshire

THE COUNTY RED DATA BOOK

Edited by
TREVOR ELKINGTON and ALAN WILLMOT

Drawings by
JACQUELINE FARRAND, STAN DOBSON, ROD DUNN
and FRED HARRISON

Foreword by Sir David Attenborough

DERBYSHIRE WILDLIFE TRUST

1996

ACKNOWLEDGEMENTS

The editors of this book would like to thank the following:

The authors for their time and co-operation in writing these chapters

Diane Wilson for her painstaking word processing of all the chapters
and preparation of the indexes and Alec Rapkin for proof reading

J. M. Tatler & Son Ltd. for their co-operation in printing this book

The many naturalists whose records form the basis of this book.

Cover photographs by Trevor Elkington (pyramidal orchid),
Fred Harrison (dark green fritillary and small blue butterflies)
and Derek Whiteley (grass snake, mountain hare, goshawk and roe deer)

FINANCIAL SUPPORT

The Derbyshire Wildlife Trust and the Editors
wish to acknowledge the generous financial support of
The Midland Bank plc, Derbyshire County Council
and the Peak Park Joint Planning Board
which has enabled the publication of this book

ISBN 1 871444 01 2

Published by the
Derbyshire Wildlife Trust
Elvaston Castle, Derby DE72 3EP

CONTENTS

FOREWORD

Sir David Attenborough

I am pleased to be associated with this work of the Derbyshire Wildlife Trust as I believe that such books can play a positive part in safeguarding our wildlife heritage.
In some ways Red Data Books make depressing reading, chronicling the result of years of neglect of our native wildlife. However, in other ways they are a sign of hope. They demonstrate the willingness of people to say this far and no further. They signal concern for what we have and they help show us what we need to do to conserve our remaining flora and fauna.

I hope all who are stewards of Derbyshire's wildlife will take heed of the problems highlighted in this book. I hope they will then work to solve the problems so that the next Derbyshire Red Data Book is a much slimmer volume. Our wildlife is a measure of the health of our environment; it needs to be cherished, not recorded and mourned in its passing.

David Attenborough

INTRODUCTION

Pat Brassley

In 1970, European Conservation Year, the Derbyshire Naturalists Trust, now the Derbyshire Wildlife Trust, hosted a conference entitled 'People and Plants in Derbyshire'. In his lecture on Rare Plants, Dr. David Shimwell put forward a series of protection measures including legislation, the establishment of nature reserves and the preparation of a 'Red Book' of Derbyshire plants.

In 1970 birds and some species of mammals were protected by Acts of Parliament, but plant protection relied on patchy by-laws which were rarely, if ever, used. There was no protection for invertebrates. The Trust was only seven years old, run by volunteers and had few nature reserves. Surveys for the national vascular plant Red Data Book had started but it was not published until 1977 and the Wild Plants and Wild Creatures Act of 1975 was still to come.

David Shimwell's Red Book was intended to include rare, precarious and decreasing species, relating to population size, changes and adverse management effects. He gave as an example the continuous local observations which had revealed the changes in populations of the precarious black mullein (*Verbascum nigrum*) in the Alport area. First recorded in 1789 by Pilkington, the largest populations were found on open lead mine spoil heaps and the smallest on stabilised areas with closed swards. Man's activities favoured this species until recently, when methods of working changed and the whole spoil heap is removed for mineral extraction.

The original idea was widened to include animals by Andrew Deadman, the Trust's first Conservation Officer, in 1976 and tentative lists were drawn up for several groups. However, it was not until 1989, that additional staff provided the opportunity to restart the project, stimulated by the production in 1988 by our partner Trust of *Endangered Wildlife in Lincolnshire and South Humberside* and the increasing number of national schemes for recording and mapping a variety of plant and animal groups.

The revised timetable provided the useful target of updating all records, positive and negative, beyond 1980 — not an easy task for the botanists faced with old, vague records but much more difficult for entomologists where recent records were sometimes few and the whole project had to start from scratch. However, the volunteer recorders took up the challenge, often getting to grips with new groups and working incredibly hard in the relatively short seasons available to many of them.

Argent and Sable Moth

A working party met regularly to provide feedback, to review the format of the proposed publication and most importantly to maintain enthusiasm. As a result of the early discussions it was decided that, bearing in mind the varying degrees of detailed information and the difficulty of surveying some groups, each group's author should determine the criteria for inclusion and should provide an explanation in their section, rather than choosing a single, arbitrary criterion to cover all groups of plants and animals which then proved unworkable for some groups. It was also agreed to include as many groups as possible, depending on the information available, in order to achieve the main aims of the book, that is to document the rare or threatened wildlife of the county and to stimulate a desire to protect those plants and animals.

Therefore, this book has had a prolonged gestation, but its appearance is now very timely for a number of reasons. Internationally there has been widespread recognition of the need to preserve and enhance biological diversity. This culminated in the 1992 Earth Summit in Rio de Janeiro which drew up the Convention on Biological Diversity, now ratified in 114 countries (including the European Union). The U.K. Government produced an Action Plan in 1994 which committed it to conserve and enhance biological diversity and in 1995 the Wildlife Trusts and other conservation groups produced a report with detailed species and habitat action plans.

In Derbyshire the last decade has seen the development of the protection of Wildlife Sites in the County, following the survey of the county outside the Peak Park in 1982-3, which formed the Wildlife Habitat Assessment. In the early 1990s the adoption of a nationally accepted computer package by The Wildlife Trusts has further increased the value of the data collected and has increasingly led to its acceptance by planning authorities as an input to the consideration of planning applications.

The information in this book presents a fascinating record, although it is sometimes disheartening to see how damaging man's activities can be. Behind the sections on each group of organisms are tens of thousands of records not only of the species listed but of those which, happily, were not included because they were found in too many sites. The wealth of data from this project combined with the new technology will assist with the integration of records for plants, animals and sites to enable appropriate conservation measures to be adopted. The surveyors and section authors hope that it will achieve the necessary conservation of threatened plants and animals but perhaps more importantly encourage others to take an interest in individual species or groups, to survey them and to understand their ecology better.

Working Party

Chairman Dr Alan Willmot

Convener Pat Brassley

Members Eileen Thorpe, Ailsa Lee, Fred Harrison, Roy Branson, Bill Grange

REFERENCES

GRIME J. P. ed. (1970) *People and Plants.* Report of a study conference. Derbyshire Naturalists Trust.

PERRING F. H. & FARRELL L. (1977) *British Red Data Book 1 Vascular Plants.* Society for the Promotion of Nature Conservation. Lincoln.

SMITH A. E. ed. (1988) *Endangered Wildlife in Lincolnshire and South Humberside: A Red Data Report.* Lincolnshire and South Humberside Trust for Nature Conservation.

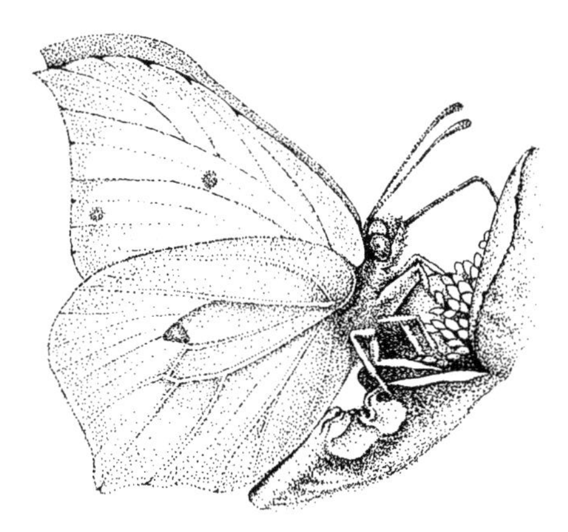

Brimstone Butterfly

HABITAT LOSS AND CREATION

WILLIAM M. GRANGE

PRESSURES ON THE DERBYSHIRE COUNTRYSIDE

Increased intensification and productivity of farming following the Second World War, encouraged by massive subsidies, has meant the maximising of land for agriculture at the expense of wildlife habitat. This, together with the use of pesticides and inorganic fertilisers, has caused drastic inroads into the wild species which had managed to adapt to the formerly less intensively managed farms. In some respects, Derbyshire has been less changed in this way than some other counties of the Midlands and eastern England where the intensification of arable farming has been on an unprecedented scale. The preponderance of pastoral farming and the fact that a substantial area of our county lies within a National Park has, to a large extent, been responsible for this. Even so, anyone who has closely observed almost any area of Derbyshire countryside over recent years would have noticed a deterioration in the richness and variety of its wildlife as a result of agricultural developments.

The post-war building boom, now in a phase of renewed activity, has eaten up huge areas of countryside — there is scarcely a village or town that has not expanded hugely in the last fifty years. A parallel increase in industrial development has also destroyed wildlife habitats and caused pollution. At the time of writing, a gigantic industrial complex has been constructed to the south-west of Derby, destroying farmland that had been run on pre-war traditional lines, rich in meadows, ponds, marshes and hedgerows (Findlay, 1990), a development which has been actively encouraged by local authorities which spend much time proclaiming their commitment and concern for the 'natural environment'.

Derbyshire is a county especially affected by extractive industries. Along the border with Nottinghamshire, the coalfield has, in the last ten years or so, seen a switch from traditional deep-mining to opencast. The old-style pits, although blighting huge areas of land with waste-tips (many of which have matured into interesting wild-life habitats) were far less damaging than the gigantic chasms of opencast mining extending over many hectares. Although, by law, opencast sites have to be 'restored' after working, the complex ecology of the former mature countryside with its old trees and consisting of

Hawfinch

an intricate pattern of macro and micro-habitats, which has taken many hundreds of years to evolve, is impossible to recreate.

Along the River Trent and the lower reaches of the Derwent in the south of Derbyshire, rich gravel deposits have long been worked. The resulting flooded pits, although they have meant the destruction of habitats such as riverside meadow and hedgerows have, in many cases, following the cessation of working, matured into valuable wildlife refuges.

The heart of the Peak District — the limestone plateau — is particularly affected by quarrying. Old, long-abandoned quarries offer both interesting geological and botanical sites — and indeed have been the source of the building stone of the delightful villages and towns of the region. The newer working quarries are a different matter. For one thing their main function is the production of road metal for an ever-expanding national road and motorway building programme; and they are on a vast scale. Whole hillsides have been quarried away and sections of dales destroyed, particularly around the towns of Buxton and Wirksworth. These huge new quarries also have reduced potential for natural colonisation when compared to the small older ones.

The resurgence in road-building has caused further losses to wildlife. For example, the new Stoke-Derby link road across southern Derbyshire, has destroyed part of a Wildlife Trust Reserve (which is also a Site of Special Scientific Interest). By-pass schemes for towns and villages, although of obvious benefit to their residents, mean that yet more open country is dissected into smaller compartments, making the movements of wild animals difficult. Motorways and similar major road schemes have not been all bad news for nature in that the wide and infrequently mown verges and embankments have become rich in wildflower species and shelter many small mammals, prey to the kestrels which are often seen hovering above them. A proposal to widen the M1 into an eight-lane highway obviously jeopardises this unexpected wildlife bonus, until the new verges so created have time to mature.

In an area like the Peak District, famed for its scenic beauty and ringed by large centres of population, it would be surprising if the large numbers of resulting visitors did not pose some environmental problems. The vast majority of people arrive by car and do not stray very far from their vehicles. Pollution from car exhausts must have an effect on the flora of roadside verges, though this is more difficult to quantify than the physical damage resulting from parking on verges. Walkers in numbers have caused harm by trampling, leading to local erosion — though in the total sum of environmental damage this is perhaps rather insignificant, though well-publicised! Popular areas, such as the Pennine Way route from Edale, and 'honeypots' such as Dovedale have, however, suffered considerable damage in this way, necessitating expensive remedial work. Walkers may sometimes disturb reptiles, ground-nesting birds, and birds of prey on the open moorlands.

The growing sport of rock-climbing has an impact too; not so much on the bare gritstone edges, but certainly on the limestone faces of the dale-sides, where rare flowering plants and ferns are put at risk by a minority of irresponsible climbers. Nesting birds of prey can also be disturbed. Caving and potholing has occasionally led to damage of unique geological features causing disturbance to roosting bats and reducing the suitability of potential refuges for larger mammals. More recently, other, less peaceful, sports have become the rage and these offer greater threats to wildlife. Moto-cross when held in woodland and other 'marginal' farmland areas causes considerable disturbance and erosion. Casual misuse by motorbikes in such locations is a growing problem. The latest craze of 'War Games' which requires woodland for its pursuit is an alarming trend.

Epidemic diseases have had a great impact on the countryside. Human agency was responsible for the introduction of a particularly virulent strain of Dutch Elm disease in the 1960s, through the importation of infected foreign timber. Most of the mature English Elms in lowland Derbyshire have been wiped out, formerly an impressive element in the landscape and host to some specialist insects, such as the White Letter Hairstreak Butterfly and several attractive species of moths — all now very scarce. The felling and clearing of most of the dead trees has eliminated any conservation benefit which may have come from the disease!

In woodlands in mid and northern Derbyshire, Wych Elm is, however, regenerating. Myxomatosis, also introduced by human agency, decimated the rabbit population during the 1950s and is still active. In some places a decline in the rabbit population has been beneficial, while, on the other hand, resulting encroachment of scrub has been detrimental to grassland communities with their wealth of flower and insect species.

THREATS TO SPECIFIC HABITATS

Woodland

The great primeval forest was, even by Neolithic times, reduced to a fragmented state through human agency. Today tree cover is very sparse indeed, consisting of scattered woodlands and copses within a largely intensively farmed landscape. The original forest is represented by a small number of 'ancient' woods which can be dated at least as far back as the Middle Ages. These comprise 'ancient semi-natural woodland', those which do not obviously originate from planting and have a history dating back, at least, to 1600; and formerly ancient woods which have been converted to plantation. Much of the existing tree-cover has, however, been established in comparatively recent times. Landowners in the 19th century organised an unprecedented planting programme and the Forestry Commission and water boards established the huge conifer plantations in the Upper Derwent and its tributary valleys in the early years of the 20th century. Derbyshire is, in spite of this, one of the least wooded of counties — approximately 5% of the land area is wooded, compared to a figure of 8% for Britain as a whole.

In wildlife terms the ancient woodlands are rated the highest — reflected in the fact that 16% of these in the County, 53% in the White Peak, have some kind of official nature conservation status — are either Sites of Special Scientific Interest, National Nature Reserves or County Trust reserves (Whitbread, 1984). Most of these can be placed in the following categories: Oak-dominated woods on fertile lowland soils, sessile oak/birch woodlands on the gritstone of the Peak District, and secondary ancient ash woods on the flanks of the limestone dales. Wet alder carr of the valley bottoms was once a widespread, and very rich, habitat and is now reduced to a few stands. These broad descriptions do not, however, convey how great is the variety in tree species composition of semi-natural woodland stands still to be found in Derbyshire.

Deciduous woods, including some ancient ones, are under threat to varying degrees. In the past they were a vital source of timber, fencing material and fuel, usually maintained by the coppice-with-standards system; a man-managed regime which itself was beneficial to a rich wildlife by creating the varied light conditions required by a range of herbaceous plants. Today many woods are neglected and overgrown. A few are used to shelter pheasant or are valued as shelter-belts. Otherwise the decline in the role of woodlands in the rural economy has made them more vulnerable to agricultural expansion. From 1920 to 1984 approximately 40% of ancient semi-natural woodland had been lost in Derbyshire (2,588 hectares), while ancient woodland converted to plantation suffered a 70% loss over the same period (Whitbread,1984).

In the Peak District almost all of this clearance was due to agricultural practice, much of it through 'grazing out', i.e. stock feeding within the unfenced woods and preventing regeneration by destroying the young trees. Eighteenth and nineteenth century industrial exploitation of the limestone dales led to a tremendous clearance of the woodlands (already adversely affected by intensive grazing by sheep) in order to provide the vast amounts of fuel required by the lead smelters. Subsequent regeneration, however, has produced species-rich secondary ancient woodland, dominated by ash. In the lowlands, destruction of the woodlands through extractive industries (opencast coal mining and gravel extraction) and urban expansion has been significant (24% and 14% respectively of the total cleared), but agriculture has claimed the rest.

Although much of the conifer afforestation in Derbyshire has been at the expense of moorland in the upper Derwent catchment area, there has been a certain amount of coniferisation of ancient woodland (e.g. at Repton Shrubs in the south), with some non-native broadleaf also planted. Although such woods still retain a high conservation value when compared to those newly planted, replanting certainly severely reduces any scientific interest.

Several non-native species of tree have invaded woodlands through self-seeding. Included here is the sycamore (*Acer pseudoplatanus*), a species native to central and southern Europe, thought to have been brought to Britain in the 16th century. Its high rate of seed fertility and establishment, coupled to very rapid growth following germination means that natives (even the equally fast-growing ash and wych elm), which are much richer in host insect species, are rapidly suppressed. In many woods on acidic soils, such as on the Millstone Grit and Bunter (Sherwood) Sandstone, the introduced Common Rhododendron (*Rhododendron ponticum*), has run riot. The dense, all-year-round shade cast by this shrub has destroyed large areas of ground-flora and, again, suppresses the regeneration of native trees and the growth of herbs.

Hedgerows

Hedgerows, though occasionally found in the Peak District, are essentially a feature of the lowlands — the eastern coalfields and the southern half of the County. A number of hedges are of great antiquity and can be traced back to mediaeval times as boundary markers, but most date from the 18th century enclosure movement. The old adage that the greater the number of species of tree and shrub present in a given length of hedge, the greater its age, is generally true.

Although loss of this man-made, but important habitat, has been greatest in the arable lands of southern Derbyshire, everywhere there has been hedge removal to enlarge fields in order to accommodate modern cultivation machinery and to maximise crop production. Road widening schemes have also taken their toll. Many surviving hedges have deteriorated as wildlife habitats due to the decline in traditional hedge-management techniques, including laying. The modern method of trimming hedges with mechanical flail cutters causes much mutilation and has been said to encourage the entry of fungal diseases through the scarified twig ends, but this has been difficult to demonstrate. Certainly, such cutting, if carried out in spring and early summer, causes considerable disturbance to nesting birds. The practice has also led to a decline in hedgerow trees because, without special care, young saplings are not allowed to grow into maturity. Figures of hedgerow loss in Derbyshire are not available, but nationally it has been estimated that, on average, about 4,500 miles of hedge per year have been removed over the last forty years. The rate of destruction has slowed considerably since 1970, following the withdrawing of grants given specifically for the removal of hedges.

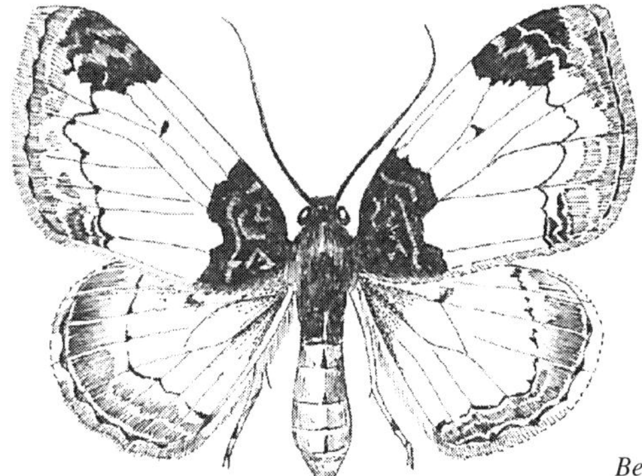

Beautiful Carpet Moth

Grasslands

The old-style flowery hay meadows of the lowlands are now almost non-existent. During the last thirty years, the practice of feeding hay to stock over the winter months has been largely abandoned, to be replaced by the use of silage. The old meadows have been ploughed up and re-seeded with cultivars of one or two species of highly nutritious grasses suitable for silage-making. Species-rich pastures have also been ploughed and re-seeded with fewer species, or have received heavy applications of inorganic fertilisers (thus preventing the growth of many wild herbs), or completely converted to arable.

On the limestone plateau of the Peak District (formerly covered by forest, and then — after its clearance — by heathland up to the beginning of the 18th century) grassland is now the dominant habitat. The vast majority of this is 'improved' cattle pasture; i.e. treated with several applications of artificial fertiliser and re-seeded with 'productive' seed mixtures. Such pastures contain about three or four grass species, with invasive herbs such as dandelion, common daisy and ribwort plantain. In recent years some fields on the high grasslands have even been ploughed out for arable, often with profits only equal to the subsidies paid. Consequently, flower-rich meadows are now rare on the limestone uplands, being found in a few enclosed fields and along the wide verges of the enclosure roads. The greatest areas of botanically rich meadow are to be found on the slopes of the limestone dales. Here are to be found some highly characteristic species, growing on land which was formerly wooded, and now an open habitat maintained by a delicately balanced grazing regime. Over-grazing by rabbits, before myxomatosis decimated them in the 1950s, led to a decrease in the number of flower species. Conversely, reduction in grazing, both by sheep and rabbits, since then, has led to a change in the species composition and an encroachment by scrub in some areas, threatening the botanical richness of the grasslands again!

Limestone quarrying has increased greatly in recent years, taking away further areas of dale-side vegetation. On the other hand old, worked out quarries and their associated spoil-heaps have in many cases developed an interesting secondary flora, especially rich in orchids. Unfortunately, such sites are sometimes singled out for the disposal of the ever-increasing mountain of domestic and industrial waste.

Along unimproved roadside verges, requirements of safety involve a mowing regime. Unfortunately this is often taken to excess, carried out as much for reasons of misguided 'tidiness' as for safety. Flail-mowing, rather than cutting, leaves a mat of detritus to smother small herbs. It must be stressed that a sensitive mowing regime is essential if a verge is to retain a floral diversity. Otherwise a few species of coarse grasses and herbs will smother the more delicate plants. One or two cuts a year, possibly a little more for the front strip of the verge, is usually sufficient.

Railway routes, both in use and abandoned, preserve interesting grassland habitats. Special mention should be made of the High Peak and Tissington Trails in the White Peak. These cycle/walk ways, converted from former railway lines, are bordered by several linear meadows of such botanical richness that they have been designated as County Trust reserves. Elsewhere, though, cuttings along old rail routes have been used for, or are at risk from being turned into, waste-disposal sites. Without active management many disused lines become invaded with scrub and lose much of their wildlife interest.

Aquatic Habitats

Still Water

Derbyshire is not noted for extensive natural bodies of water. Human activity has, however, created a number of reservoirs, flooded gravel pits and ornamental lakes, fish, mill and farm ponds.

The big reservoirs of the upper Derwent are nutrient-poor and have too great a fluctuation in water level to be of exceptional value in wildlife terms. These great reservoirs, and the one recently completed at Carsington in the southern Peak District have, in fact, taken away more than they provide in that they have destroyed former woodland, meadow and marsh in the valley bottoms. Smaller reservoirs tend to be more useful habitats, especially for wetland and water birds, though recreational pressures in the form of sailing, wind-surfing, power-boating, bank and boat fishing, etc, can be highly detrimental. At Staunton Harold Reservoir in south Derbyshire S.S.S.I. status has been withdrawn because of the drastic reduction in the bird population through such water sports. Nature conservation interests were disappointed to learn that the new Carsington Reservoir is to be developed specifically to cater for water-sports.

The string of gravel pits bordering the River Trent and lower Derwent in south Derbyshire contain some interesting wetland habitats, especially those which have had time to settle down following the end of extraction operations. They are good places for small forms of aquatic life, for fish, breeding birds, marginal aquatic vegetation, and scrubland. Recreational pressures have, again, reduced the wildlife value of some gravel pits. Others are under threat from infill by domestic waste, followed by 'landscaping' and 'restoration'. Canals, especially when disused, form secluded linear ponds of great richness, to which recreational pressures and insensitive dredging operations are destructive. Some have been infilled or drained for reasons of safety.

Due to the ban on the use of static water bodies for watering stock there is no practical reason for the retention and maintenance of farm ponds. Many have consequently been filled in, choked by vegetation — part of a natural succession leading to marsh, bog or willow carr (which in practice usually quickly dries out and is subsequently grazed) — or destroyed by building developments or quarrying. On the limestone plateau, often the only bodies of water for miles around are the wholly artificial circular dew ponds. Although looking very unpromising as wildlife habitats, they are sometimes vital breeding sites for amphibians. Alas, many former dew ponds are waterless because of cracking of their waterproof linings.

From a survey covering 113 adjacent one-kilometre squares in the limestone area of the Peak District — utilising a combination of map and field data — it has been deduced that there was a loss of 43% in the number of ponds from 1899 to 1974, with a further loss of 33% over the fifteen years to 1989 (Monk, pers comm). This demonstrates what a frightening acceleration there has been in the loss of a very special habitat. Surviving ponds are under threat from nutrient-rich farm run-off, causing over-enrichment (eutrophication) and hence oxygen depletion.

Marshes, Bogs, and Lowland Fens

Apart from the presence of 'flashes' caused by subsidence in mining areas, marshlands are now of small extent and rare in lowland Derbyshire — farm drainage schemes have seen to that. Falling ground-water levels (a national problem), due to a combination of drought and increased water consumption, pose a serious threat to those which have survived.

Marshes and wet flushes, however, still occur abundantly on the hillsides in the Dark Peak, with small ill-drained hollows at lower levels, and extensive expanses of *Sphagnum* bog on the moorland tops. In places on the gritstone lime-rich flushes and seepages occur, creating conditions for a very distinctive fauna and flora. *Sphagnum* mosses are extremely vulnerable to air pollution, particularly to sulphur dioxide, produced in quantity from the industrial conurbations which press in closely on both the west and east flanks of the northern moorlands, falling as acid rain. Lead pollution has also been indicated in the decline of the *Sphagnum.* Although lead mining has ceased, limestone quarrying in the White Peak may create a lead-rich dust which is carried by prevailing winds over the adjacent moors of the Dark Peak (Anderson & Shimwell, 1981). Over the years *Sphagnum* bog has decreased drastically, while cotton grass bog, much poorer in associated species, has increased at its expense. Air pollution may be ultimately responsible for the tremendous erosion which has taken place on the summit plateau of many of the moorland areas, notably on Kinder Scout, the highest part of the Peak District, where huge expanses of bare peat, totally devoid of vegetation, are dissected by ever-deepening and widening gullies.

Rivers and Streams

In general, the major rivers of Derbyshire have not been so drastically treated as some in other parts of the country. This applies particularly to the Derwent, whose banks exhibit a remarkable intactness and natural interest along much of its length, not only through its Peak District reaches, but also within the city of Derby. South-east of the City centre, however, large-scale industry impinges on the river. In recent years accidental pollution incidents have caused the large-scale death of fish, but in general the river is relatively clean.

In complete contrast to the Derwent are the River Rother and its tributary, the Doe Lea, in the north-east of Derbyshire. Both rivers are grossly polluted, being devoid of aquatic life over much of their courses. Other rivers and streams in the lowlands, including the Trent and Erewash, are also susceptible to both industrial and farm pollution (cattle slurry, inorganic fertilisers and pesticides) and to domestic sewage. In the Peak District, the Derwent, Dove, Wye and other rivers, being surrounded by low intensive pasture, are not, in general, subject to serious agricultural pollution. One problem affecting the Derwent, the Trent, and other rivers, especially in the lowlands, is the recent spread of two introduced flowering plant species. The annual Himalayan Balsam, (*Impatiens glandulifera*) is a spectacularly beautiful plant with hooded pink flowers — native to the Himalayan region. The perennial Japanese Knotweed, (*Fallopia japonica*), although it dies back to each rootstock in the winter, throws up a huge shrubby growth in spring. Both these large and robust species blanket long reaches of riverbank, suppressing native vegetation. They have proved very difficult to control.

Some streams have been canalised as part of drainage and flood-prevention measures, creating a grossly simplified habitat. Certainly, enough disturbance to our water-courses must have been perpetrated to have caused the apparent extinction of breeding otters in our County.

Heather Moorland

Heather-dominated moor occupies the slopes of the hills in the Dark Peak, below the bogs on the flatter tops. It has long been managed to support grouse, though varying numbers of sheep are also present. In order to encourage the regeneration of new shoots required by the grouse, heather moor is burnt on a rotational basis, following a 12 to 15 year cycle. Although this treatment, when carefully carried out, is beneficial to the heather in the long term, it is obviously locally catastrophic to some insects and reptiles. Conversely, the bare areas initially produced by burning have been shown to be important for some groups such as ground beetles and spiders (Gardener & Usher, 1989). However, apart from bilberry and a few grasses, burning is detrimental to most associated plant species. Burnt moor is therefore poorer in plant species than unburnt, leading to increased grazing pressure by sheep on the heather itself. Where sheep density is high, heather moor has reverted to poor grassland, often with invasion by bracken. It is significant that there has been a three-fold increase in the number of sheep on the Peak District moors since the 1930s, though this trend might well be drastically reversed if subsidies are withdrawn. Burning of moorland is now strictly controlled by legislation and should not cause any further loss of the heather. Accidental fires are another matter. In recent very dry summers, many hectares of moor have been seriously damaged in this way. This is another undesirable effect of the increased use of the countryside for recreation.

Other destructive pressures on moorland have been the establishment of conifer plantations — notably around the big northern reservoirs, and on Matlock Moor in the south-eastern Peak District. However, certain insects, and birds such as the nightjar have benefited from the new habitat created by the mix of conifer and heather moorland at the margins of the plantations.

Between 1913 and 1980 there has been a drastic loss in heather moorland in Derbyshire — ranging from 33% for heather-dominated moor to 77% for heather and mixed grass type. Bilberry-dominated moorland has suffered a 46% loss.

The Wormwood Moth

Specialised Man-made Habitats

Buildings, especially farm out-buildings and barns can provide roost sites for bats and nest sites for birds. Reduction in the numbers of the rare Barn Owl has undoubtedly been accelerated by the replacement of old brick, stone and wooden structures by metal buildings, and the conversion of old barns into dwellings. Roof spaces in houses, both old and new, are excellent bat-roosting sites, but wood preservation chemicals used in such locations have been lethal to these animals. Old mine-workings and railway tunnels function as artificial caves. These are liable to be filled in for safety reasons, or capped. Awareness of bat conservation and recent legislation has, however, led to the installation of bat grilles at a few sites, but more need to be provided.

Colliery waste tips, of which there are a great many in eastern and southern Derbyshire, when allowed to mature undisturbed, develop an interesting flora, while lagoons impounded by them are attractive to migratory waders. Tips are generally regarded as being unsightly and are often 'landscaped', grassed or forested, thus reducing their potential diversity. Fortunately, wherever possible, valuable wildlife habitats on these sites are now protected during reclamation work.

Churchyards often contain a meadow-type vegetation which is a relic of what the now intensively farmed surrounding fields used to be like. Unfortunately they are especially vulnerable to the efforts of a particularly strong tidy-up lobby and much work is carried out by church authorities and volunteers in mowing and manicuring away any botanical (and entomological) interest. Often this zeal is carried as far as scrubbing lichens from the tombstones!

CREATION OF NEW HABITATS

Reclamation of colliery spoil heaps, and of other industrial sites has already been mentioned, as well as other sites inadvertently becoming new wildlife habitats, such as motorway verges. Not all of these, however, have nature conservation as a top priority. Where this is ostensibly the case, as with restoration schemes following opencast coal-mining, creation of new habitats cannot compensate (although they are so promoted) for the intricacies and specialised ecological niches of the semi-natural habitats of the former countryside. This must be borne in mind when an industrial concern uses 'restoration' and 'creation of new habitats' in its arguments in pressing for a particular development, which will destroy long-established wildlife sites.

Nevertheless, in the wake of the increasing concern felt at the losses to our wildlife, there is a movement under way in the general community to deliberately create new habitats. These have, however, usually been on a small scale such as wildlife gardens, and nature reserves in school grounds. Apart from the case of the perhaps now ill-named Common Frog, whose survival has probably owed much to garden ponds, such small reserves, created from scratch, are rarely of direct benefit to endangered species. They may well be so as they proliferate, mature and develop. They certainly do perform a useful role in promoting the merits of nature conservation and allow people, who otherwise rarely have the opportunity, to come into close contact with nature. Their educational role is, of course, of immense long-term benefit to wildlife conservation.

The recently well publicised 'Midlands Forest', to be created under the aegis of the Countryside Commission, is to extend from N.W. Leicestershire to East Staffordshire across the southern part of Derbyshire, an exciting prospect for nature conservation. However it is a project which will require considerable finance, commitment and co-operation from land-owning interests and farmers. It will also require considerable expertise in wildlife conservation from those carrying out the planning and supervising. Blanket forestry, even if carried out in native broadleaf species would be detrimental to some valued open wildlife habitats such as grasslands. Much current tree planting, although carried out with the best intentions, is in some very inappropriate locations!

In the meantime there is much scope for the creation of new habitats throughout the county. Vast areas of public parks are still managed to create the 'green desert' of mown grassland, with a few scattered trees. Instead, they could support wildflower meadows, miniature woodlands and natural-looking ponds and lakes and this is being carried out by some local authorities.

The farm 'set-aside' policy could be organised more positively for nature conservation, with farmers receiving an income in return for the active management of land for wildlife, alongside their normal agricultural activities as in Environmentally Sensitive Areas (ESAs) — given the necessary training, help and advice from statutory and non-statutory conservation organisations. This would certainly be a healthy change from the former dubious policy of paying farmers for not destroying sites of wildlife value. The farmer's role would therefore come to be seen as being very different to what it is now — more that of a general custodian of the countryside, rather than only that of a producer of food at a profit (with subsidies being paid for massive surpluses) to whom wildlife and wild places are a dispensable impediment!

REFERENCES

ANDERSON P. and SHIMWELL D. (1981) *Wild Flowers and other Plants of the Peak District.* Moorland Publishing Co., Ashbourne

FINDLAY, A. (1990) Burnaston: A Relic of the English Countryside — A Future for the Car Industry? *Derby Natural History Society, Observations No. 16*

GARDENER, S. M. and USHER, M. B. (1989) Insect Abundance on Burned and Cut Calluna Heath. *The Entomologist* 108: 149-157

WHITBREAD, A. (1984) Derbyshire: *Inventory of Ancient Woodland.* Nature Conservancy Council

General Reading:

Derbyshire County Council/Derbyshire Wildlife Trust (1985) *Derbyshire Wildlife Habitat Assessment - Report of Survey*

ELKINGTON, T. T. ed. (1986) *The Nature of Derbyshire.* Barracuda Books, Buckingham.

Forestry Commission. *Census of Woodlands and Trees (Derbyshire), 1979 - 1982*

HARRISON F. & STERLING M. J. (1985) *Butterflies and Moths of Derbyshire* Part 1. Derbyshire Entomological Society

Holly Blue Butterfly

FUNGI

TONY LYON

Fungi are a distinctive group of microscopic organisms allotted their own Kingdom (Mycota). They are predominantly filamentous and obtain their nutrients by absorption from their surroundings. They may be free-living, in which case they live by rotting dead plant or animal tissues and feeding off the products of their decomposition, or they may live in association with other living organisms. Such associations include mutually beneficial ones such as lichens (fungi with algae or cyanobacteria) and mycorrhizas (fungi with plant roots) and those where the fungus lives at the expense or another organism which derives no benefit from the partnership (fungal diseases of plants and animals).

Concern for the conservation of fungi has lagged behind that for plants and animals. This almost certainly reflects the interests of conservationists in the different groups of organisms, but is also due to the relative paucity of information about the ecology, distribution and frequency of fungal species. Only recently has a provisional red data list of British fungi been published (Ing, 1992). The list 'attempts to identify those British fungi which are extinct or at serious risk of becoming extinct because their habitats, substrata or hosts are similarly threatened'. The levels of risk are expressed in IUCN categories.

Ex Extinct or probably extinct. Species which have not been found for several decades, usually not in the present century.

E Endangered. Species occupying habitats which are disappearing fast or those occurring in such small populations or in so few sites as to make their extinction likely in the near future.

V Vulnerable. Species likely to become endangered in the near future if the causal agents of their decline are not removed or reduced.

R Rare. Species with small populations or few sites which are not at present threatened but may become at risk without adequate protection.

This account includes any species in the provisional red data list for which the author has been able to find records for Derbyshire. Sources of information which have been searched include published species lists from fungus forays of the British Mycological Society and the Yorkshire Naturalists Union, and a database containing information about fungi recorded over fourteen years by the author, the Sorby Natural History Society of Sheffield, the Sheffield Fungus Research Group and the Derbyshire Fungus Study Group.

Fungus recording in Derbyshire has been restricted to a relatively limited number of sites, many of which lie in the northern part of the county. The majority of records are of 'larger fungi', those species whose fruitbodies are big enough to be seen without the use of a microscope, in particular the agarics (or toadstools). Agaric fruitbodies usually persist for only a few days, so that the chances of their going unrecorded are high. It is probable that such fungi are seriously under-recorded.

The principal threats to fungi are from loss of woodland or grassland habitats due to changes in forestry or agricultural management and from pollution. Woodland management for fungal diversity includes leaving dead wood on site to provide substrata for decomposers. Diverse communities may be found in planted woodland types that do not occur naturally in the county — mature beechwoods (*Fagus sylvatica*) for example support a particularly rich fungal flora. Grasslands that show greatest fungal diversity are those of low productivity on nutrient-poor soils. These are found on both the Carboniferous limestone and the acidic rocks of the Millstone Grit and Coal Measures and are a special feature of north Derbyshire. They are at risk from agricultural improvement by fertilisers.

Studies in continental Europe, for example Holland, show that pollution, particularly by nitrogen compounds, may have a marked influence on fungi. Mycorrhizal species appear to be specially susceptible. Given the levels of industrial and agricultural activity in and around Derbyshire, pollution may be having a significant impact upon fungi in the county. Unfortunately systematic recording of fungi in Derbyshire does not extend sufficiently far back for us to be able to recognise any significant changes in their occurrence.

Hygrocybe calyptraeformis

	National status	*No. of sites*
Ascomycotina		
Graddonia coracina On fallen twigs in damp places	R	3
Basidiomycotina		
Camarophyllus atropunctus In grazed nutrient-poor grassland	R	1
Cortinarius violaceus Mycorrhizal associate of deciduous trees, particularly oak, beech and hornbeam	E	1
Hygrocybe calyptraeformis In grazed nutrient-poor grassland	V	4

	National status	*No. of sites*
Marasmius hudsonii On fallen holly leaves	R	1
Mycena rubromarginata On rotting coniferous tree-stumps, logs and twigs	V	1
Pseudocraterellus sinuosus Among litter in deciduous woods, particularly beech	V	1
Ripartites metrodii In leaf litter under deciduous trees	R	1
Russula carminea Mycorrhizal associate of deciduous trees, especially birch	R	1
Strobilomyces strobilaceus Mycorrhizal associate of both coniferous and deciduous trees; Derbyshire records are with beech.	V	2

REFERENCES

ING, B. (1992) A provisional red data list of British fungi. *The Mycologist* 6: 124-128

Strobilomyces strobilaceus

LICHENS

Oliver Gilbert

The lichens of Derbyshire are reasonably well known. A flora was published in 1969 (Hawksworth 1969); since then there have been four supplements (Hawksworth 1974; Gilbert 1983; Gilbert 1993; Gilbert & Ardron 1995). The north of the county has been more thoroughly investigated than the south.

During the modern period (post-1960) 427 lichens have been recorded from Derbyshire. Selection for inclusion in the Red Data Book has been made from this list. An additional 70 species were known to the older lichenologists. Many of these, particularly the epiphytes, are believed to have become extinct due to past high levels of air pollution; others have been eliminated by landuse change, while in a few cases there is uncertainty about the exact identity or locality of the species.

Selection criteria

About half the species recorded during the modern period are known from at least ten localities, many being common and widespread throughout the county. A high proportion of those with less than ten extant records are under recorded because of their small size and a requirement for identification to be confirmed using a microscope, e.g. many species on limestone. In addition an increasing number of lichens belonging to genera such as *Cladonia, Lecidea, Lepraria* and *Porpidia* need 'chemical finger printing' using thin layer chromatography to separate them from closely related taxa; this is expensive, time consuming and not often done. A further group is under recorded because they occur in poorly worked habitats e.g. urban wasteland.

The balance of species which have a genuinely restricted distribution in Derbyshire can be further reduced by removing a group of conspicuous epiphytes which are currently spreading throughout the North Midlands in response to declining levels of sulphur dioxide air pollution. It includes *Parmelia caperata, P. perlata, P. revoluta, P. subrudecta, Physcia aipolia* and *Usnea subfloridana*. Before the end of the century this group is expected to have become widespread.

The remaining 105 lichens are considered to have a truly restricted distribution in the county as a result of specialised habitat requirements or because they are near to the edge of their geographical range. Several are national rarities, e.g. *Absconditella sphagnorum, Lecidea commaculans* and *L. pernigra*; appear to have their main centre of distribution in the UK in Derbyshire, e.g. *Lecanora campestris* subsp. *dolomitica* (type locality Creswell) and *Protoparmelia picea*; or represent outlying populations, e.g. *Micarea pycnidiophora* (otherwise known only from a small group of woodlands in southern England, *Mycoblastus alpinus* (Chatsworth Park is the only site outside Scotland). With regard to habitat, the species are almost evenly split between epiphytes (28%), the limestone (24%) and the gritstone moors (24%). Igneous rock outcrops (8%), streams (6%) and disused lead rakes (4%) also have their specialities.

Legislation

A national Red Data Book for lichens is in preparation, and includes three species currently known in Derbyshire. These are marked with an asterisk. A number of lower plants are protected under Schedule 8 of the Wildlife and Countryside Act 1981. None of the 16 lichens are currently known from Derbyshire though there are old records of the elm epiphyte *Caloplaca luteoalba.*

Throughout this account nomenclature follows The Lichen Flora of Great Britain and Ireland (Purvis *et al.*1992). The number of localities from which each species is known is indicated.

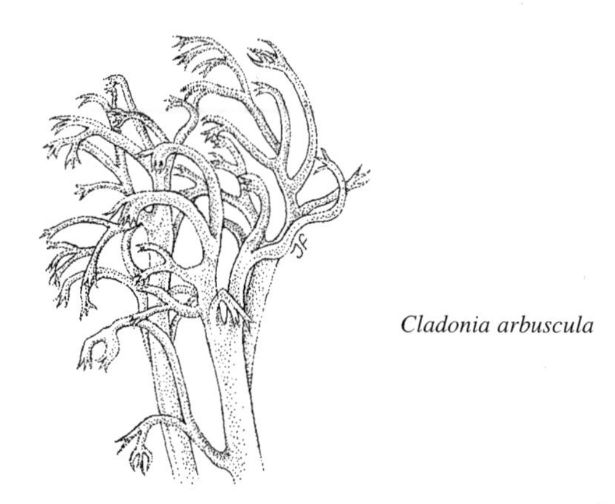

Cladonia arbuscula

Post-1960 Localities

***Absconditella sphagnorum** 1

Very rare, on *Sphagnum fimbriatum*, Upper Padley.

Acarospora glaucocarpa 1

Rare, sheltered limestone, Cave Dale, Castleton.

Amygdalaria pelobotryon 3

In very small amount on igneous rock outcrops at sites in the Tideswell area and near Castleton.

Anaptychia ciliaris 4

Several patches among moss on top of limestone walls in Perry Dale, at Peak Forest and near Pilsbury. A remarkable occurrence of a species which is highly sensitive to air pollution.

Arthonia punctiformis 2

Very rare, on the smooth bark of ash and rowan, Monsal Dale and by the Highlow Brook. Last seen 1973.

A. radiata 1

Very rare on the smooth bark of ash, Monsal Dale. Last seen 1963.

Arthopyrenia lapponina 1

On the smooth bark of rowan, Highlow Brook. Last seen 1973.

Post-1960 Localities

A. punctiformis 4
Rare, on the smooth bark of ash and rowan.

Bacidia arceutina 1
Occasional on sheltered, vertical Magnesian Limestone at Creswell Crags.

B. fuscovirens 4
On damp, shaded limestone by rivers and at the base of cliffs.

B. rubella 3
Very rare on mossy, dust impregnated tree boles in the bottom of limestone dales.

B. viridescens 3
A terricolous species known from a few lead rakes; possibly overlooked.

Baeomyces roseus 3
Known from a few sites on acid moorland and a disused railway track.

Bryoria fuscescens 4
Small relic populations survive on ancient oaks and sheltered sandstone boulders in the Old Deer Park, Chatsworth and at Rowtor Rocks, Birchover. Also on an old rowan in Hangman's Clough, Goyt's Moss.

Buellia ocellata 1
A small population on igneous rocks, Masson Hill.

Calicium glaucellum 3
On a few ancient oaks in the Old Deer Park, Chatsworth, near Stoke Ford and in the Goyt Valley.

Caloplaca arenaria 2
Small populations on igneous rock at Carlton Hill Quarry, Chelmorton and in Taddington Dale.

C. cerina 1
On the twigs of dust impregnated ash trees, Stoney Middleton.

C. chalybaea 4
Limestone outcrops in Dove Dale, Lathkill Dale and Monks Dale.

Candelariella xanthostigma 1
On elms in Stanton Deer Park. Last seen 1973, possibly extinct.

Post-1960 Localities

Cetraria islandica 5

This northern species, here at its southern limit in the Pennines, is now known from four sites on the limestone heath (Longstone Moor, Parwich Moor, Coombs Dale and above Bradwell). A new site on Millstone Grit near Barbrook Reservoir, Big Moor was discovered in 1990. All populations are small.

Chaenotheca brunneola 1

On decorticated oak, Beeley Moor.

Chaenotheca stemonea 1

On the dry undersides of a few ancient oaks, the Old Deer Park, Chatsworth.

Chaenothecopsis cf. nigra 1

On the dry underside of a single ancient oak, the Old Deer Park, Chatsworth.

Cladonia arbuscula 3

Once considered extinct in the county, this species has been refound on the limestone heath at Longstone Moor, Parwich Moor and near Sheldon.

Cladonia bellidiflora 1

On heather thatched roof of church porch, Old Brampton. This upland lichen may be an introduction at this site.

C. caespiticia 1

Scarce on peat, Ramsley Moor, west of Curbar.

C. ciliata *var.* **tenuis** 4

Known from sites on the limestone heath, most are associated with mining disturbance.

C. digitata 2

On rotting oaks in the Old Deer Park, Chatsworth and Padley Gorge.

C. foliacea 1

In open calcareous grassland, Hassop Common.

C. fragilissima 7

Scarce, acid grassland in Cote Clough, Oyster Clough, above Upper Padley and in a few additional sites. Something of a local speciality.

C. luteoalba 15

Associated with most major outcrops of the Millstone Grit. This species is extremely rare on a world scale; it probably has its headquarters in the Pennines.

Post-1960
Localities

C. parasitica 2

On a single tree stump, the Old Deer Park, Chatsworth and in Park Wood Nature Reserve, Taxal.

Collema limosum 2

On compacted basic soil in urban areas. Whitwell churchyard and a demolition site in Glossop.

Collema polycarpon 1

On limestone in Monsal Dale; last seen 1973.

Coriscium viride 1

Once considered extinct, rediscovered on peat in Padley Gorge (1988).

Cornicularia normoerica 1

A single thallus of this rare arctic-alpine is present at Horse Stones, Upper Derwent.

Dermatocarpon luridum 2

Many records appear to be based on misidentifications but reliably reported from wet rocks in Beresford Dale and near Topley Pike.

Farnoldia jurana 2

On limestone in Dovedale and Conies Dale.

Fuscidea austera 3

On gritstone boulders in the Upper Derwent. An upland species.

F. kochiana 3

Occasional on sandstone boulders, Combs Edge and Cracken Edge Quarries.

Graphis scripta 2

Poorly developed on hawthorn in Millers Dale. Last seen 1965. Recent records from the Goyt Valley.

Hymenelia prevostii 2

On carboniferous limestone in Coombs Dale and Monks Dale.

Hypocenomyce caradocensis 2

On the bark and lignum of old oaks. Frequent in the Old Deer Park, Chatsworth and at Stake Side in the Errwood Valley.

Post-1960
Localities

Ionapsis epulotica 1

On limestone in Dovedale. Last recorded 1964.

Lecania cuprea 1

On limestone outcrop, Coombs Dale. Possibly overlooked.

L. inundata 1

One patch on igneous rocks near Tideswell. This species is not well known.

L. rabenhorstii 1

Sheltered limestone cliff, Deepdale, Sheldon. Possibly overlooked.

L. sylvestris 1

Sheltered limestone cliff, Coombs Dale. A poorly known species.

Lecanora campestris *subsp.* **dolomitica** 12

Known from several localities on the Magnesian Limestone where this endemic species appears to have its headquarters. Most records are from walls. Also one record from dust impregnated bark and three from Carboniferous limestone walls in the Peak District. Type locality: Creswell Crags.

L. pallida 1

Dust impregnated bark of ash trees, Churn Hole, Topley Pike.

L. piniperda 1

This upland species is known from the stump of a single oak in the Old Deer Park, Chatsworth.

***Lecidea commaculans** 1

On gritstone boulder, Burbage Moor. Otherwise only known from two sites in the Scottish Highlands.

L. diducens 1

Dry gritstone stone wall, Upper Derwent.

L. hypopta 2

Occasional on the lignum of fallen oaks in the main park and the Old Deer Park, Chatsworth.

L. lactea 1

On a single igneous boulder in a field near Peak Forest.

Post-1960 Localities

L. pernigra 1

The first British record (1979) was from Cracken Edge Quarries near Chinley where the species is frequent. Also on Kinder Scout. Since discovered to be abundant just over the county boundary at Laddow Rocks, Crowden Great Brook, Cheshire.

Lecidella elaeochroma 1

On sheltered trees in Dovedale and Lathkill Dale. Last seen 1965.

Lempholemma chalazanellum 1

On a mortared wall near Creswell Crags.

L. myriococcum 1

On shaded limestone in Lathkill Dale. Rare.

Leptogium diffractum 2

Sheltered limestone in two dales.

L. subtile 1

Shaded limestone boulder in the river, Cresbrook Dale.

L. teretiusculum 3

Scarce on old lead spoil, Haydale and Bradwell Moor.

Micarea pycnidiophora 1

On the base of an oak, Limb Valley. Previously known only from a few old woodlands in southern England.

Mniacea jungermanniae 1

Lynch Clough, Upper Derwent.

Mycoblastus alpinus 2

On sheltered sandstone boulders in the upper part of the Old Deer Park, Chatsworth and near Baslow. First English record (1991) of this predominantly Scottish species.

Ochrolechia frigida 1

Rare, Hurkling Stones; southern-most record in the Pennines.

O. tartarea 4

Small populations of this northern species are known on boulders at Burbage Edge, below Ladybower Tor and at the north end of Chatsworth Park.

Post-1960 Localities

Parmelia acetabulum 1

On elm, Stanton Deer Park. When last seen (1972) the species was in poor condition and is probably now extinct.

P. conspersa 1

A single large thallus on a sandstone roof flag, Beeley Village.

P. elegantula 1

On the dust impregnated boles of several sycamores, Chelmorton.

P. exasperatula 1

On the capstones of a limestone wall near Arbor Low.

P. pastilifera 2

On the capstones of limestone walls under trees near Arbor Low and near Brassington.

P. tiliacea 5

Rare on deciduous trees at five well scattered localities.

Parmeliopsis aleurites 1

Occasional on the lignum of fallen oaks, the Old Deer Park, Chatsworth.

Peltigera horizontalis 1

On a mossy igneous outcrop near Tideswell.

P. leucophlebia 4

On mossy limestone outcrops at several sites in Back Dale, Deep Dale and Bradford Dale.

***P. ponojensis** 1

On lead spoil, Eyam. Possibly an overlooked species.

Pertusaria pupillaris 1

On a single oak by the Highlow Brook below Stoke Ford. Not seen since its discovery in 1973.

Placynthium garovaglii 5

Dry limestone underhangs on major outcrops.

Placynthium tantaleum 3

Rare on limestone in three dales.

Post-1960 Localities

Polyblastia cruenta 2

In two streams in the Upper Derwent.

Polysporina dubia 1

On mortared gritstone retaining wall near Ashopton Viaduct.

Porpidia hydrophila 2

Beside moorland streams, rare: Slippery Stones and Kinder Downfall.

Protoparmelia oleagnia 1

The Old Deer Park, Chatsworth. A rare species of ancient woodland.

Protoparmelia picea 7

This nationally local species has recently been recognised as abundant on the gritstone edges of the Eastern Moors.

Psora decipiens 1

Disused quarry near Chee Tor, Millers Dale. Seen once in 1965.

Ramalina fraxinea 2

A few thalli on dust contaminated ash trees down wind of Hope Valley Cement Works. Also on a poplar near Chapel-en-le-Frith.

Rhizocarpon subgeminatum 5

On the gritstone edges of the Eastern Moors. An upland species.

Rinodina roboris 1

Very rare on trees in Dovedale. Last seen 1964.

Sarcogyne privigna 2

On mortared gritstone walls at Ashopton and Stoney Middleton. This species is declining nationally.

Schaereria cinereorufa 3

This northern upland species is locally abundant on igneous boulders near Peak Forest and on certain gritstone edges.

Solorina spongiosa 4

Rare, calcareous ledges in Back Dale, Cave Dale and covering a quarry floor at Hartington.

Post-1960 Localities

Stereocaulon dactylophyllum 3

Rare, on walls at Slippery Stones and near Foolow, also a shooting butt on Bamford Moor.

S. nanodes 5

Known from six sites; three on mine spoil, two on igneous rock outcrops, also in Monyash churchyard.

S. vesuvianum *var* **symphycheiloides** 1

Occasional, disused railway sidings at Willington Power Station.

Strangospora moriformis 1

On an ash tree at Fernilee, near Whaley Bridge.

Thelidium pluvium 1

On wet rocks, Cote Clough, Derwent Reservoir.

Trapeliopsis glaucolepidea 4

Local, on wet peat.

Umbilicaria deusta 3

Frequent on damp boulders along 3km of the Burbage Brook above Padley Wood; also by a track near the Ladybower Inn.

U. polyrrhiza 5

Boulders on high moors, mostly in the Derwent area.

U. torrefacta 2

Small populations of this northern species, here at its southern limit in the Pennines, occur on boulders at Gardom's Edge and the north end of Padley Wood.

Usnea filipendula 2

Abundant on a single boulder in oak woodland at two localities: Ladybower Wood and the Old Deer Park, Chatsworth. This species has almost certainly survived the industrial period at these sites.

Verrucaria caerulea 3

On sheltered limestone, Cave Dale and Gratton Dale.

V. laetebrosa 1

On limestone boulder in the winterbourne, Lower Cressbrook Dale.

REFERENCES

GILBERT, O. L. (1983) The lichen flora of Derbyshire — Supplement 2. *Naturalist, Hull* 108: 131-137

GILBERT, O. L. (1993) The lichen flora of Derbyshire — Supplement 3. *Naturalist, Hull* 118: 3-8

GILBERT, O. L. & ARDRON, P. A. (1993) [1995] New, rare and interesting lichens from North Derbyshire. *Sorby Record* 30: 48-53

HAWKSWORTH, D. L. (1969) The lichen flora of Derbyshire. *Lichenologist* 4: 105-193

HAWKSWORTH, D. L. (1974) The lichen flora of Derbyshire — Supplement 1. *Naturalist, Hull*, 99: 57-64

PURVIS, O. W., COPPINS, B. J., HAWKSWORTH, D. L., JAMES, P. W. and MOORE, D. M. (1992) *The Lichen Flora of Great Britain and Ireland*. London: Natural History Museum.

Ramalina fraxinea

INDEX
Scientific Names (Genera)

MOSSES AND LIVERWORTS

Tom Blockeel

The Bryophyta are a group of green spore-bearing plants. They include the true mosses (Musci), the liverworts (Hepaticae), and some smaller groups including bog mosses (*Sphagnum*) and hornworts (Anthocerotae). Most bryophytes develop leafy shoots, but some liverworts are thalloid (i.e. they form flattened, often branched structures which are not differentiated into stems and leaves). They are small plants, the majority of species having shoots in the range 5mm to 10cm long.

Bryophytes differ greatly from the ferns and their allies in the absence of vascular tissue and in their reproductive biology. The male and female 'inflorescences' of bryophytes are borne on the leafy shoots or thalli. The fertilised egg gives rise to a sporophyte which remains attached to and dependent on the parent plant. The sporophyte takes the form of a capsule, often raised on an elongate stalk (or seta). Spore-bearing tissue is contained within the capsule, and the ripe spores are released at maturity by a variety of mechanisms.

Bryophytes are sometimes confused with lichens and even algae, neither of which produce leafy shoots. Lichens are fungal in origin and any resemblance to bryophytes is entirely superficial. Superficially similar lichens may be separated from thalloid liverworts by their leathery texture and differently coloured upper and lower surfaces.

It is a common perception that bryophytes are plants of damp habitats, but this is only a partial truth. Certainly the most luxuriant communities are found in damp woods and by rocky streams, but bryophytes occur in all manner of terrestrial habitats and some are adapted to survive conditions of prolonged heat and drought. They are important colonisers of bare soil, rock and bark, and may assume dominance in some specialised habitats (e.g. peat bogs).

National Legislation and Red Data Book

The British Isles have a rich bryophyte flora, and communities of international importance are well represented in northern and western areas. Thirty-three species are legally protected in the United Kingdom, having been added to Schedule 8 of the Wildlife and Countryside Act 1981. One of the listed species (*Thamnobryum angustifolium*) occurs in Derbyshire, and is in fact unknown elsewhere.

A national Bryophyte Red Data Book is currently in preparation.

Derbyshire data and limitations

The only complete flora of Derbyshire bryophytes was published by Linton (1903), who brought together the large number of records which had accumulated in the latter part of the nineteenth century. During the present century, there have been relatively few workers in the county, and there is no up-to-date and comprehensive account of the flora. General accounts covering parts of the county have been published by Adams (1956), Hall (1962) and Furness & Lee (1985). These works mention few specific stations and are written from an ecological rather than a distributional standpoint.

Many records for Derbyshire have been accumulated during the preparation of the Atlas of British Bryophytes (Hill, Preston & Smith, 1991-94), some of them compiled during excursions of the British Bryological Society. It had been hoped that the Atlas database might be available in time for the preparation of this account, but unfortunately this has not been possible.

The present account is therefore based largely on my own records from the county, which date from 1985. These are supplemented by records kindly provided by other bryologists with knowledge of the county, and with data from published excursion accounts (Clarke, 1970; Hill, 1973; Corley, 1976). Other unpublished reports and lists exist, but these are not always adequately reliable. A cautious approach has therefore been adopted in citing the numbers of known stations.

Some 500 bryophyte species have been recorded to date from Derbyshire, not including subspecies and varieties. Recent recording has added significantly to this total, but there are also 61 species for which I have not been able to verify any recent (post-1970) records. These are listed in Table 1. It should not be assumed that all these species have been lost from the county; some of them undoubtedly still occur and will be refound. The true number of extant species in the county is likely to exceed 460.

Selection Criteria

Because of the incomplete nature of the data available at this time, it has not been possible to use entirely objective criteria in compiling the list of threatened and vulnerable species.

The basic criterion used is that the species should have no more than six known extant stations in the county. An extant station is deemed to be one recorded since 1970. This stringent criterion reflects the relatively poor state of our present knowledge of the Flora. For the same reason it has been necessary to exercise some subjective judgement in applying it.

(a) Some critical and/or inconspicuous species are thought to be significantly under-recorded and have been excluded if they are widespread and not threatened at the national level.

(b) A few species with more than six extant stations have been included if (i) knowledge of their distribution in the county is thought to be fairly complete, and (ii) they are nationally scarce species or they habitually occur in small quantity.

It is unlikely that many species have been omitted from the list which ought to have been included, but more complete recording will probably justify the removal of some of them.

Threats

Bryophytes, no less than other organisms, have been and continue to be affected by the destruction and degradation of natural habitats. There are, however, a number of factors which have a particular relevance to bryophytes.

Perhaps the most significant of these is atmospheric pollution. Epiphytic bryophytes and species of dry acid rock surfaces have been severely affected by air-borne pollutants, including SO_2. Species of dry well-illuminated trees (for example in the genus *Orthotrichum*) are more affected than those of mature closed woodland. Similarly the absence of bryophytes on the surfaces of dry grit rocks, particularly on the higher moorland crags, is noteworthy in many parts of north Derbyshire. Species of genera such as *Andreaea*, *Hedwigia* and *Grimmia* are much less widespread than previously, and in some cases have completely disappeared. This effect is less apparent on the Carboniferous limestone, partly because the limestone rocks are further removed from the sources of pollution, but also because the calcareous content of the rock itself acts as a buffer against the pollutants.

It is possible that the increased acidity of habitats caused by acid rain, rather than the toxicity of individual pollutants, has been responsible for the demise of some species. Some of the epiphytic species which have been reduced or eliminated (e.g. *Orthotrichum* spp., *Cryphaea heteromalla*) require a relatively base-rich bark; conversely, certain species thought to have increased during the present century (e.g. *Dicranoweisia cirrata*, *Dicranum tauricum*) are acidiphile.

It is also widely accepted that the degradation of the South Pennine blanket bog, and in particular the loss of *Sphagnum* species, has been due in large part to pollutants such as sulphur dioxide. Burning of the moors and overgrazing have been contributory factors. Healthy bogs support a wide range of *Sphagnum* species and associated bryophytes, including a number of characteristic leafy liverworts. However all the bogs of the southern Pennines are degraded to a greater or lesser extent, and many of the characteristic bog bryophytes, including some *Sphagnum* species, have been eliminated. In spite of reductions in the levels of SO_2 pollution, there is little evidence of any immediate recovery in the bog flora.

There is, conversely, evidence that some base-rich niches have disappeared from the gritstone moors. There are nineteenth century records from the Kinder district of *Breutelia chrysocoma*, a species now confined to the limestone dales, and *Scorpidium scorpioides*, a species of base-rich flushes. The apparent loss of such species may be a further indication of increased acidification.

Grazing at sustainable levels is often beneficial to bryophytes in maintaining open habitats. This is particularly so on thin calcareous soils, as in the limestone dales. Interesting communities of bryophytes may be found in short turf on thin or stony soil, and on limestone ledges. Some of these are ephemeral species which flourish during the autumn and winter when the rainfall is highest. Certain species which occur on south-facing crags and slopes have a southerly distribution in the British Isles and are nationally rare or scarce. These species are intolerant of shade, nor can they compete with tall herbs.

This problem also affects old limestone quarries. Communities occur in soil pockets on rock ledges, stony ground and spoil heaps which require an open habitat and invariably disappear with the invasion of coarse grass and scrub. At least one nationally rare species (*Lophozia perssonii*) is confined in Derbyshire to this habitat.

A different habitat which supports interesting communities of short-lived bryophytes is provided by fallow arable land. These communities are of interest ecologically because of the different strategies which the species must adopt for survival in a temporary habitat. Some species are able to complete their life cycle and produce mature spores in a matter of weeks; others have specialised means of vegetative propagation. Intensive agricultural practices have reduced the diversity and extent of these communities, which depend for their development on the land lying fallow for sufficient time after harvesting.

Current changes in the flora

The most noteworthy trend in recent recording has been the increasing number of records of epiphytic species. These records provide strong evidence that some species (e.g. *Orthotrichum pulchellum* and *Cryphaea heteromalla*) are regaining lost ground following recent reductions in the severity of SO_2 pollution. The effect has been most noticeable in humid sites, but has been apparent also in drier places in the limestone dales. This is not a reason for complacency, however; air pollution remains a significant factor in limiting the distribution of many bryophytes.

TABLE 1: Species not recorded since 1970

The occurrence in the county of the species enclosed in square brackets requires confirmation. They are species which for various reasons have been misunderstood and misreported in the past.

Hepatics

Moerckia hibernica
[Lophozia longiflora]
Plagiochila spinulosa
Douinia ovata
Scapania compacta
Odontoschisma sphagni
Cephaloziella stellulifera
Kurzia pauciflora
Lepidozia pearsonii

Mosses

[Sphagnum molle]
Sphagnum tenellum
Polytrichum alpestre
Pogonatum nanum
Buxbaumia aphylla
[Seligeria calcarea]
Dichodontium flavescens
Dicranella crispa
Campylopus atrovirens
Fissidens limbatus
Aloina brevirostris
Aloina rigida
[Desmatodon convolutus]
Pterygoneurum ovatum
Pterygoneurum lamellatum
Acaulon muticum
[Gymnostomum recurvirostrum]
Weissia squarrosa
Weissia rutilans
Weissia rostellata
Weissia longifolia var. angustifolia
Schistidium alpicola
Schistidium strictum
Grimmia orbicularis
Dryptodon patens
Racomitrium ericoides
Campylostelium saxicola
[Orthodontium gracile]
Bryum uliginosum
Bryum intermedium
Bryum creberrimum
Bryum pallescens
Plagiomnium elatum
Amblyodon dealbatus
Philonotis caespitosa
Philonotis calcarea
Bartramia hallerana
Orthotrichum lyellii
Orthotrichum rivulare
Orthotrichum stramineum
Orthotrichum tenellum
Hedwigia ciliata
Antitrichia curtipendula
Neckera pumila
Pterigyandrum filiforme
Drepanocladus sendtneri
[Drepanocladus vernicosus]
Scorpidium scorpioides
Scleropodium cespitans
Rhynchostegium megapolitanum
Homomallium incurvatum
Rhytidiadelphus subpinnatus

TABLE 2: Additional species

The following species have been excluded from the primary list but are noted here because they belong to one of the following categories: (a) they are scarce in Derbyshire but do not quite meet the criteria defined for the primary list; (b) they are nationally scarce but have good populations in Derbyshire and are not threatened locally; or (c) they are potential candidates for the primary list but are thought to be significantly under-represented in recent records and are therefore inadequately known.

Hepatics

Apometzgeria pubescens
Leicolea alpestris
Tritomaria quinquedentata
Jungermannia obovata
Scapania aspera
Cephaloziella hampeana
Ptilidium ciliare
Porella cordaeana
Frullania dilatata
Frullania tamarisci
Lejeunea cavifolia

Mosses

Sphagnum capillifolium
Tetrodontium brownianum
Polytrichum longisetum
Atrichum crispum
Pleuridium acuminatum
Pleuridium subulatum
Ditrichum flexicaule
Ditrichum crispatissimum
Dicranum majus
Dicranum bonjeanii
Fissidens crassipes
Fissidens osmundoides
Encalypta vulgaris
Pottia intermedia
Weissia microstoma
Discelium nudum
Funaria muhlenbergii
Pohlia cruda
Bryum gemmiferum
Amphidium mougeotii
Fontinalis squamosa
Hookeria lucens
Sanionia uncinata
Calliergon cordifolium
Brachythecium albicans
Taxiphyllum wissgrillii

The nomenclature of species follows Hill, Preston & Smith (1991-94).

For each species the number of recent records is cited. The notation 'n+n' is used in cases where the Atlas of the Bryophytes of Britain and Ireland (Hill, Preston & Smith,1991-94) shows a wider distribution in the county than that encompassed by the known records. The first figure is the number of known extant stations; the second is the number of additional 10-km squares for which a modern record is mapped in the Atlas.

Pedinophyllum interruptum

Hepatics (Liverworts)

Targionia hypophylla 7

A Mediterranean/Atlantic species occurring on sunny usually south-facing rocks in the limestone dales. Threatened by scrub encroachment. The Derbyshire population of this species is of national importance.

Preissia quadrata 7+1

On damp limestone rocks in the limestone dales, on natural outcrops and also in old rock cuttings, usually at higher altitudes. Not immediately threatened.

Riccia cavernosa 2

On damp mud at the edge of reservoirs. Variable in occurrence from year to year and dependent on the exposure of mud flats in summer as water levels recede.

Riccia fluitans 2

An aquatic of still water in lakes and ponds. Threatened by water pollution but able to withstand some eutrophication.

Riccia subbifurca 1

On exposed sandy/peaty soil on a reservoir margin. Very rare. Threatened because of the small size of the population, and dependent on the exposure of bare ground in summer as water levels recede.

Riccia glauca 4

On bare soil. Recent records are all from fallow fields on the Coal Measures. Threatened by intensive agricultural practices.

Metzgeria fruticulosa 6

An epiphytic species, most commonly on Elder bushes, but susceptible to atmospheric pollution. However, recent records suggest that the species may be regaining lost ground as levels of SO_2 pollution decrease.

Metzgeria temperata 2

An epiphytic species, on tree boles and branches in humid sites. Susceptible to atmospheric pollution, and with a more westerly distribution in Britain than *M. fruticulosa.*

Metzgeria conjugata 7

On moist shaded rocks, including Millstone Grit. Possibly under-recorded because of confusion with *M. furcata.*

Blasia pusilla 1

On wet shale in a moorland clough. The species occurs on bare moist soil on tracks and banks in other parts of Britain, but has apparently not been recorded recently from this habitat in Derbyshire.

Fossombronia incurva 1

On moist sandy soil in an old Silica pit on the limestone plateau. A rare but easily overlooked species, the Derbyshire station being isolated from the nearest known localities elsewhere in England. Threatened by possible in-filling.

Fossombronia wondraczekii 4

An ephemeral species of damp soil in fields and on peaty ground by reservoirs, but probably under-recorded.

Barbilophozia barbata 3+1

Among grit rocks on the moors where there is some mineral enrichment, and in stony turf (possibly leached) in a limestone dale. Vulnerable because of the small size of the populations.

Lophozia sudetica 2

Confined to high ground, among grit rocks on moors where there is slight base enrichment. Vulnerable because of the small size of the populations.

Lophozia excisa 4

On gritty track-sides, on lead-mine spoil, in old Silica pits, and on leached soil on rock ledges in the limestone dales. Probably under-recorded.

Lophozia perssonii 2

On calcareous spoil in old limestone quarries. This is a nationally rare species. Its conservation is difficult, as it is threatened both by the re-opening of old quarries, and by natural plant succession in abandoned ones.

Lophozia incisa 5+1

On wet rocks in moorland cloughs, especially on wet shale. Rare and usually in small quantity, and therefore vulnerable.

Lophozia bicrenata 3

On peaty and gravelly soil and thin soil over grit rocks. Apparently rare, but inconspicuous and possibly under-recorded.

Leiocolea badensis 3

Crevices of limestone rocks and on limestone spoil. Apparently rare, but distribution uncertain because of confusion with the related *L. turbinata*, and probably not immediately threatened.

Sphenolobus minutus 1

On the sheltered sides of Millstone Grit boulders in block scree on moorland. Very rare and vulnerable because of the small size of the population.

Tritomaria exsectiformis 6

On sheltered Millstone Grit boulders, mostly in Oak/Birch woodland.

Mylia taylorii 6

In damp block scree on the high moors, very rarely on heathy banks at lower altitudes.

Mylia anomala 1

On *Sphagnum* and peat in bogs. Very rare and almost exterminated by pollution and degradation of the peat bogs.

Jungermannia exsertifolia 5

In wet stony flushes on Millstone Grit moorland. Plentiful in one of its stations and not immediately threatened.

Jungermannia caespiticia 3

On moist gritty or shaly soil on moorland banks. A nationally rare species, recently discovered on the northern moors. A colonist of exposed soil and erratic in occurrence, but probably not threatened.

Jungermannia hyalina 1

On grit rocks by wooded streams. Rare, but inadequately known. Threatened by pollution of streams.

Jungermannia paroica 3

On moist grit boulders on wooded stream banks.

Nardia geoscyphus 1

On peaty soil. Very rare but probably under-recorded.

Pedinophyllum interruptum 1

On moist, shaded limestone in woodland. This species is almost confined in Britain to the Pennine limestones. The known site is in a Reserve.

Lophocolea fragrans 1

In crevices of grit boulders on a river bank. The single Derbyshire station is of great interest, being geographically isolated from the main centre of distribution of this species in Britain, which is along the western coasts. There is only a small quantity at the Derbyshire station and the species is therefore endangered.

Scapania cuspiduligera 1

On thin soil on grit rocks influenced by basic seepage. Very rare, and threatened because of the small size of the population.

Scapania scandica 6+1

On gritty soil and on the sides of sheltered Millstone Grit boulders. Probably under-recorded.

Scapania lingulata 1

On moist grit rock with slightly basic seepage. The Derbyshire station, which was discovered in 1990, is the only one known in England, but the species is a critical one and probably under-recorded.

Scapania irrigua 3+3

On damp sandy and gravely soil. Apparently rare but probably under-recorded.

Scapania umbrosa 5

On the sides of sheltered grit boulders in woodland and block scree.

Scapania gracilis 3+1

On grit boulders in Oak woodland, rare and confined to the older, mature woods.

Odontoschisma denudatum 1

On rotting logs. One extant station, which lies within the part of Derbyshire formerly in Cheshire.

Cephaloziella rubella 1

On peat under *Calluna.* Known from one station only but the species is critical and almost certainly under-recorded.

Cephalozia lunulifolia 4+2

On boulders and wood in acid woodland. Scattered localities on the Millstone Grit.

Cephalozia connivens **1**

On a moist gritstone boulder by a spring in Oak woodland. More commonly found in *Sphagnum* bogs and on rotting wood, but not confirmed recently from these habitats in Derbyshire.

Nowellia curvifolia 2+1

On rotten logs in damp woods. Threatened by the removal of logs in otherwise suitable woodland.

Cladopodiella fluitans 2

In very wet bogs. Has almost disappeared as a result of the degradation of the moorland bogs.

Hygrobiella laxifolia 5

On wet cliffs in moorland cloughs where there is slight mineral enrichment. Very local and confined to the Kinder/Bleaklow area. The Derbyshire populations are at the south-eastern limit of the distribution of this species in Britain.

Kurzia trichoclados 4

On damp grit rocks in block scree on moorland and in Oak woodland. Rare but plentiful in one of its stations.

Bazzania trilobata 4

Among grit boulders in block scree on moorland and in Oak woodland. Rare and in small quantity in its known stations.

Calypogeia integristipula 2

On gritty soil on crags and at the base of boulders in Oak woodland. A critical and probably under-recorded species.

Calypogeia trichomanis 3

On wet peaty/gritty soil in moorland cloughs. Apparently rare but probably not threatened.

Blepharostoma trichophyllum 1

The recent record is from slightly base-rich gritstone rocks in woodland; recorded previously from calcareous flushes and likely still to be present in this habitat.

Trichocolea tomentella 3

On wet grit rocks and in calcareous marshes. In very small quantity at all its known sites and therefore endangered. Threatened by water pollution and damage to the fragile flush habitats in which it grows.

Ptilidium pulcherrimum 4

On trees and shrubs in woodland on both limestone and Millstone Grit.

Radula complanata 3+1

On shaded limestone and as an epiphyte, very rare. Sensitive to atmospheric pollution, but possibly benefiting from fall-out from limestone quarries.

Porella arboris-vitae 1+1

Among sheltered Carboniferous limestone rocks. Very rare.

Marchesinia mackaii 4

On sheltered limestone cliffs, mostly on the Carboniferous limestone, but with one station on the Magnesian limestone. A Mediterranean/Atlantic species, nearly confined to the west of Britain. The Derbyshire stations are of interest as some of the very few for the species in central England.

Lejeunea lamacerina 7

On wet grit rocks along streams. A few scattered localities on the Millstone Grit, chiefly in the Derwent Valley.

Cololejeunea calcarea 3+1

On moist shaded limestone. Scattered localities in the limestone dales.

Cololejeunea rossettiana 6+1

On shaded limestone, often in drier sites than *C. calcarea*. Scattered localities in the limestone dales.

Mosses

Sphagnum teres 2

In enriched seepages and flushes in moorland districts.

Sphagnum girgensohnii 3

On moist moorland banks and in Oak woodland. Apparently very rare, but possibly overlooked.

Sphagnum russowii 5

On moist banks on moorland. Scattered localities on the Millstone Grit.

Sphagnum quinquefarium 2

On rocky banks in Oak woodland and among *Calluna* on moorland banks.

Andreaea rupestris 8

On Millstone Grit rocks in open woodland and moorland cloughs. Scattered localities, but always depauperate and in small quantity. Susceptible to atmospheric pollution.

Andreaea rothii 6

On Millstone Grit rocks, especially where periodically irrigated. A few scattered localities, with some healthier populations than *A. rupestris*.

Polytrichum alpinum 6

On steep banks in high moorland cloughs, usually on mineral soil. Not immediately threatened.

Diphyscium foliosum 1

Peaty soil and rock crevices. Very rare, a single recent station, which lies within the part of Derbyshire formerly in Cheshire.

Archidium alternifolium 2

On damp sandy and peaty soil, recent records being from reservoir margins.

Distichium capillaceum 3

On base-rich rocks on both Magnesian and Carboniferous limestone.

Distichium inclinatum 2

In Carboniferous limestone dales. One recent record is from an artificial habitat on old stone steps.

Brachydontium trichodes 4

On damp porous Millstone Grit in moorland cloughs. A few stations in the High Peak.

Seligeria donniana 3+1

On moist limestone on sheltered crags. Confined to the Carboniferous limestone, rare but probably under-recorded.

Seligeria pusilla 5+2

On moist limestone on sheltered crags. Confined to the Carboniferous limestone, rare but probably under-recorded.

Seligeria brevifolia 1

On damp Millstone Grit. Recently discovered in a single station in the High Peak, in very small quantity. Otherwise known in Britain only from Snowdonia and two stations in Scotland. Endangered because of the very small size of the population.

Seligeria acutifolia 4

On moist limestone on sheltered crags. Confined to the Carboniferous limestone, rare but probably under-recorded.

Seligeria trifaria 3

On wet limestone on vertical crags and in rock crevices in the limestone dales, in small quantity. The British distribution of this species is centred on the Pennine limestones. Endangered by the small size of the populations.

Rhabdoweisia crispata 1

In a wet underhang on a streamside cliff on the High Peak moors. Endangered because of the small size of the population.

Cynodontium bruntonii 3

In crevices of dry Millstone Grit rocks. Endangered because of the small size of the populations.

Dicranella subulata 1

On damp rocks on a streamside cliff on the High Peak moors.

Dicranum fuscescens 6+1

On grit boulders in Oak woodland. A few stations on the Millstone Grit.

Dicranum montanum 3

On tree boles in mature woodland. A few stations on the Carboniferous limestone.

Dicranodontium denudatum 1

On Millstone Grit boulders in woodland. Apparently very rare, but possibly overlooked.

Campylopus fragilis 1

On base-poor soil, including leached turf on ledges of Carboniferous limestone crags. Probably under-recorded but also confused with related species.

Leucobryum glaucum 5

In Oak woodland, on moorland and in limestone heath. Rare, plentiful only at one station. Eliminated from most moorland areas by burning and overgrazing.

Fissidens rivularis 2

On stones in wooded streams on Millstone Grit. A species of Mediterranean/Atlantic distribution, known in two isolated stations in Derbyshire, long distant from all other British localities. Threatened by water pollution.

Fissidens exilis 0+4

On bare moist soil on sheltered banks and in woodland.

Fissidens celticus 1

On damp soil on a wooded stream bank on the Coal Measures. The Derbyshire station is an isolated occurrence of a species which has a southern and western distribution in the British Isles.

Tortula princeps 2

On limestone crags and ledges. A nationally rare species.

Tortula laevipila 1

An epiphytic species, susceptible to atmospheric pollution.

Tortula virescens 2

An epiphytic species, recently discovered in two stations in the county.

Tortula marginata 5

On shaded basic rock and walls. Scattered stations, principally on the Magnesian limestone. Not immediately threatened.

Aloina aloides 8

On open calcareous soil, nearly always in old limestone quarries and cuttings.

Pottia starkeana *sensu lato* 3

An ephemeral species of thin, bare calcareous soil. Probably under-recorded. Threatened by agricultural improvement on the Magnesian limestone and by scrub invasion in the limestone dales.

Pottia lanceolata 6

On bare calcareous soil, principally in the limestone dales. Threatened by scrub invasion.

Pottia bryoides 2+1

On open calcareous soil among limestone rocks.

Pottia recta 5

An ephemeral species of calcareous soil. A few stations on Carboniferous and Magnesian limestone. Possibly overlooked because of its small size and seasonal occurrence.

Phascum curvicolle 1

An ephemeral species of calcareous soil. Apparently very rare but possibly overlooked because of its small size and ephemeral occurrence. Threatened by scrub invasion and agricultural improvements.

Barbula acuta 1

On bare calcareous ground. A somewhat critical species, possibly overlooked but also confused with other members of the genus.

Barbula reflexa 2+1

In calcareous turf and among limestone rocks.

Barbula spadicea 1

On wet cliffs and rocks by streams.

Barbula tomaculosa 2

On clay soil on arable land on the Coal Measures. Nationally rare but only recently described and possibly overlooked.

Barbula nicholsonii 2

On stonework on a river bank, and on old asphalt. Only recently detected in Derbyshire and possibly under-recorded.

Barbula sinuosa 5+2

On calcareous rocks by streams, and on damp sheltered walls and paving. Probably much under-recorded.

Bryoerythrophyllum ferruginascens 1

On bare soil in an old Silica pit on the Carboniferous limestone plateau. Threatened by possible in-filling.

Gymnostomum aeruginosum 4+1

On moist Carboniferous limestone crags. Very local but plentiful at some of its sites.

Gymnostomum calcareum 7

On tufa and damp limestone rocks. Scattered stations on Carboniferous and Magnesian limestone. Not immediately threatened.

Anoectangium aestivum 1

On wet weakly base-rich cliffs by a moorland stream. The Derbyshire station is an isolated occurrence of this montane species at the edge of its British range.

Oxystegus tenuirostris 9

On wet grit rocks by streams, especially near waterfalls, often in Oak woods. Scattered localities on Millstone Grit, and always occurring in small quantity.

Tortella nitida 1

On Carboniferous limestone crags. Endangered because of the small size of the population.

Pleurochaete squarrosa 3

A Mediterranean/Atlantic species of stony calcareous ground. Threatened by scrub invasion.

Grimmia laevigata 1

On dry base-poor rock. The sole station is on a single igneous rock in a limestone dale, and is an isolated locality for a nationally rare and decreasing species. Endangered because of the small size of the population.

Grimmia donniana 3

On grit boulders and walls. Has probably decreased because of atmospheric pollution, especially in habitats on natural rock. On walls, the mortar probably acts as a buffer against pollutants.

Racomitrium aquaticum 3

On wet grit rocks on crags and by streams. Very rare and endangered because of the small size of the populations.

Racomitrium sudeticum 1

On a gritstone wall in the High Peak. A species of upland and montane districts at the edge of its British range in Derbyshire.

Ptychomitrium polyphyllum 4+2

On grit rocks and walls. Rare, but fairly plentiful in one or two places on mortared grit walls by reservoirs.

Funaria fascicularis 2

On soil on limestone rock ledges, probably where slightly leached.

Physcomitrium pyriforme 1+4

On mud in marshes and wet fields, mostly in the lowlands and probably therefore under-recorded.

Physcomitrium sphaericum 2

On exposed mud by reservoirs. Variable in occurrence from year to year and dependent on the exposure of mud flats in summer as water levels recede. A nationally rare species.

Physcomitrella patens 4+1

On mud, especially by reservoirs. Scattered localities, but sometimes abundant.

Ephemerum sessile 1

In pockets of damp soil on stony ground at the margin of a reservoir. This species is dependent on the exposure of bare ground in summer as water levels recede.

Tetraplodon mnioides 3

On animal remains (usually sheep) in damp or boggy ground on the high moors. Transitory in occurrence because of its habitat.

Splachnum sphaericum 4

On dung in boggy ground on moors. Very rare, but found plentifully on cow dung at one site in 1992.

Schistostega pennata 4

In crevices and holes in friable grit and sandstone rocks.

Pohlia elongata 1

On ledges of grit rocks in a moorland clough. Endangered because of the small size of the population.

Pohlia drummondii 3

On gravelly tracks and in old Silica pits. Rare but possibly over-looked.

Pohlia bulbifera 3

On damp peaty soil, especially by reservoirs.

Pohlia camptotrachela 4

On damp peaty and base-poor soil, including reservoir margins. Probably under-recorded.

Pohlia muyldermansii 3

On wet cliffs by moorland streams in the High Peak.

Pohlia lutescens 3

On sandy and loamy soil, probably under-recorded.

Pohlia lescuriana 1

On moist peaty soil by a moorland stream. Probably under-recorded.

Anomobryum filiforme 1

On a wet cliff in a moorland clough.

Plagiobryum zierii 2+2

On slightly basic rock ledges, principally on Carboniferous limestone.

Bryum algovicum 2+1

On thin soil on limestone boulders and in old quarries.

Bryum inclinatum 2

On shale banks. Probably under-recorded.

Bryum elegans 1

On Carboniferous limestone crags.

Bryum alpinum 1

Irrigated grit rocks in a moorland clough. Endangered because of the very small size of the population.

Bryum sauteri 1

On soil in a stubble field. An inconspicuous species, which is probably under-recorded.

Bryum tenuisetum 2

On bare peaty soil. An inconspicuous species, which is probably under-recorded.

Rhodobryum roseum 4+2

In thin soil on ledges on Carboniferous limestone outcrops.

Mnium thomsonii 2

On north-facing limestone crags. Only recently found in Derbyshire and possibly overlooked. A rare species of sub-montane distribution in Britain.

Mnium marginatum 2+1

Among limestone rocks in woods. Rare but possibly under-recorded.

Rhizomnium pseudopunctatum 3

In wet, base-rich flushes in moorland cloughs. Vulnerable because of its occurrence in small quantity in a scarce and fragile habitat.

Plagiopus oederianus 6

On shaded Carboniferous limestone, in woodland or on steep north-facing slopes. Scattered localities in the limestone dales.

Bartramia pomiformis 9

On ledges and crevices of grit rocks. Scattered localities, but usually in small quantity and therefore vulnerable.

Bartramia ithyphylla 3

In rock crevices in moorland cloughs. Very rare and vulnerable because of the small size of the populations.

Philonotis arnellii 3

On damp sandy/peaty soil on reservoir margins.

Breutelia chrysocoma 3

In moist limestone grassland, in small quantity. An Atlantic species, at the edge of its British range in Derbyshire. Threatened by agricultural improvements.

Zygodon baumgartneri 1

On base-rich grit in a moorland clough. More commonly epiphytic, but not known in this habitat in Derbyshire. Vulnerable because of the small size of the population.

Zygodon conoideus 2

An epiphytic species, commonly on Elder. Only recently detected in Derbyshire, and thought to be spreading in some parts of England in response to improvements in air quality.

Orthotrichum striatum 1

An epiphytic species, sensitive to atmospheric pollution.

Orthotrichum sprucei 2

On tree roots in the flood zone of rivers. Occurring in small quantity and threatened by water pollution. The Derbyshire rivers do not have a sufficiently deep and reliable flood zone for the optimum development of this species.

Orthotrichum pallens 1

An epiphytic species, on old Elder. The Derbyshire record is an isolated occurrence of a nationally rare species which is sensitive to atmospheric pollution.

Orthotrichum pulchellum 8

An epiphytic species, usually on old Elder, sensitive to atmospheric pollution. Recent records suggest that the species may be regaining lost ground as levels of SO_2 pollution decrease.

Ulota phyllantha 7

An epiphytic species, sensitive to atmospheric pollution. The species is in small quantity at each of its sites, but it has only recently been detected in Derbyshire and may be spreading in response to improved air quality.

Cryphaea heteromalla 4

An epiphytic species, most commonly on old Elder, sensitive to atmospheric pollution. Recent records suggest that the species may be regaining lost ground as levels of SO_2 pollution decrease.

Leucodon sciuroides 1

On Carboniferous limestone rocks on a dale-side. This species commonly occurs as an epiphyte on tree boles but has not been recorded recently from this habitat in Derbyshire.

Thamnobryum angustifolium 1

On limestone by a spring. The Derbyshire station is the type and only world locality for this species. A single patch only is known, and the species is endangered because of the small size of the population.

Thuidium delicatulum 2

Recent records are from an old lead mine, and a base-rich flushed ledge in a moorland clough. A critical species, confused with *T. philibertii.*

Thuidium philibertii 4

In calcareous turf. Rare but possibly under-recorded. Threatened by the spread of coarse grasses and scrub.

Thuidium recognitum 3

In thin turf and in stable Carboniferous limestone scree.

Campylium calcareum 1+1

Calcareous ground and tree bases. One recent record, but probably under-recorded.

Amblystegium fluviatile 3+2

Stones by streams. Threatened by water pollution.

Amblystegium tenax 2+2

In and by streams, including rivers in the limestone dales.

Amblystegium varium 2+1

On damp wood and stones. Rare but probably under-recorded.

Amblystegium compactum 2

On shaded but bare calcareous soil. On Carboniferous and Magnesian limestone. Plentiful at one site on the Magnesian limestone.

Platydictya confervoides 3

On damp limestone and tufa. Scattered localities on Carboniferous limestone.

Platydictya jungermannioides 2

In rock crevices and caves on wet limestone.

Drepanocladus revolvens 3+2

In slightly enriched moorland flushes.

Drepanocladus cossoni *(D. revolvens* var. *intermedius)* 1

In a base-rich moorland flush.

Drepanocladus exannulatus 5

In slightly enriched moorland flushes.

Isothecium holtii 1

On Millstone Grit boulders by a stream. An Atlantic species which is very rare in the Pennines. The Derbyshire station is geographically isolated from its nearest other stations.

Isothecium striatulum 1

On shaded Carboniferous limestone boulders in woodland. The Derbyshire station is an isolated locality for this southern species.

Brachythecium glareosum 1+2

On calcareous ground, probably under-recorded.

Brachythecium salebrosum 1

In damp woodland. Only recently recorded in Derbyshire and probably under-recorded.

Brachythecium mildeanum 2

In damp grassland and stony ground. Very rarely recorded but a critical species and possibly overlooked.

Brachythecium appleyardiae 6

On dry Carboniferous limestone rock ledges, especially at the base of sheltered crags and under overhangs. A nationally rare, recently described species now known from several sites in Derbyshire.

Rhynchostegium lusitanicum 1

On rock in swift-flowing moorland streams. An Atlantic species at the edge of its British range in Derbyshire.

Eurhynchium speciosum 1

In wet woodland.

Rhynchostegiella teesdalei 6

On wet limestone and grit rocks by shaded streams. Threatened by water pollution.

Orthothecium intricatum 6+1

In base-rich rock crevices. Very rare on Millstone Grit, scattered stations on the Carboniferous limestone. Not immediately threatened.

Entodon concinnus 2+1

In short calcareous turf. Apparently very rare, though possibly overlooked. Threatened by growth of coarse grasses and scrub.

Plagiothecium latebricola 1

Tree bases in wet woodland. Very rare, and threatened by drying out of woodland.

Plagiothecium laetum 5

On peaty soil and about tree bases. Scattered stations, and possibly under-recorded.

Isopterygium pulchellum 6

On slightly base-rich rock ledges, mostly on the Millstone Grit. Vulnerable because it is present in small quantity at most of its sites.

Pylaisia polyantha 5

An epiphytic species; very local, but plentiful in one of its stations and possibly benefiting from the fall-out of lime-dust from quarries.

Hypnum lindbergii 4

On damp peaty and gravelly soils, on reservoir margins and in thin grassland.

Rhytidium rugosum 5

In short turf and stable scree on Carboniferous limestone slopes. A few stations in the limestone dales, but very locally plentiful.

Rhytidiadelphus loreus 2+2

In old woodland, in small quantity. Apparently absent from the High Peak moors, although common on moorland in other parts of Britain. Probably eliminated over wide areas by a combination of pollution, burning and overgrazing.

Hylocomium brevirostre 2+1

On and among boulders in limestone turf and old woodland. Intolerant of disturbance to its habitat. However, one of the known sites is in a Reserve.

ACKNOWLEDGEMENTS

I am very grateful to the following individuals, who have provided records or other information for use in the preparation of this account: Dr. M. E. Newton, M. A. Pearman, A.V. Smith, R. Thomas.

REFERENCES

ADAMS, F. W. (1956). The Bryophytes of the Sheffield Region. In: *Sheffield and its Region*, ed. D.L. Linton, pp 321-334. British Association for the Advancement of Science.

CLARKE, G. C. S. (1970). The Autumn Meeting 1970 [Sheffield]. *Transactions of the British Bryological Society* 6: 391-393.

CORLEY, M. F. V. (1976). The meeting at Knutsford, Cheshire, April 1976. *Bulletin of the British Bryological Society* 28: 7-9.

FURNESS, S. & LEE, J. (1985). Mosses and Liverworts. In: *The Natural History of the Sheffield Area and the Peak District*, ed. D. Whiteley, pp 68-83. Sorby Natural History Society, Sheffield.

HALL, R. H. (1962). Flowerless Plants. In: K. C. Edwards, *The Peak District*, pp 96-107. New Naturalist, Collins, London.

HILL, M. O. (1973). The Spring Meeting 1972 [Matlock]. *Journal of Bryology* 7: 517-518.

HILL, M. O., PRESTON, C. D. & SMITH, A. J. E. (1991-94). *Atlas of the Bryophytes of Britain and Ireland.* Volume 1 *Liverworts (Hepaticae and Anthocerotae).* Volume 2 *Mosses (except Diplolepideae).* Volume 3 *Mosses (Diplolepideae).* Harley Books, Colchester.

LINTON, W. R. (1903). *Flora of Derbyshire*. Bemrose. Derby.

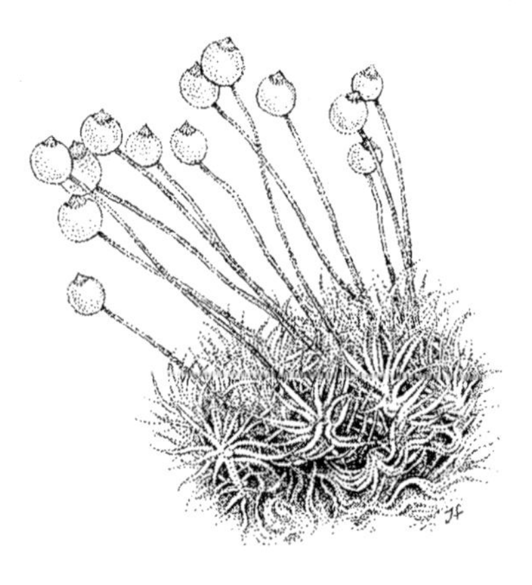

Bartramia pomiformis

INDEX
Scientific Names (Genera)

FLOWERING PLANTS, FERNS AND ALLIES

PAT BRASSLEY

This section includes most of the groups commonly thought of as plants; fungi are dealt with separately. For some groups such as stoneworts and algae there was insufficient information although the project has stimulated an interest in them. Flowering plants include not only plants with brightly coloured flowers which enhance the countryside, but also trees and shrubs, some of which have inconspicuous flowers or fruits. Grasses, sedges and rushes have less attractive, but often conspicuous flowers. Yew (*Taxus baccata*) and juniper (*Juniperus communis*) are the native representatives in Derbyshire of the conifer group, the latter only occurring in one site. The ferns and their allies include three groups of plants found in Derbyshire, the clubmosses, true ferns and horsetails.

When the project started a list was drawn up of all species considered to be native for which there were less than twenty site records at Derby Museum. This included all the records gathered for the *Flora of Derbyshire* (Clapham, 1969) and the two supplements. This resulted in a list of more than 300 species, even after the exclusion of genera such as *Rubus* (brambles) with 400 micro-species and *Hieracium* (hawkweeds) with 250 micro-species on the grounds of the difficulty of identification. The initial list was added to by local botanists. Some species are Derbyshire specialities; others are those which are rare nationally but relatively common in the county and these are listed at the end of this section before the species accounts.

A list of all the known site records for each species was compiled and circulated to botanists and then divided into 10 km squares for volunteer recorders to check. The records varied from accurate eight figure grid references to a location name or 10 km square number, the majority being four figure and not always accurate. In rechecking all the old sites, botanists were asked to improve the grid reference record and whether the plant was still present, not refound or destroyed. Location details were added where appropriate; in a narrow limestone dale it can save time to know which side and how far up the daleside the plant was recorded. Some individuals took up the challenge and the full records, held by the Wildlife Trust, show their dedication to tracking down plants in isolated and unexpected places from inadequate and erroneous grid references.

New localities were also noted but there was not time for systematic searches in new areas. Therefore there is a heavy reliance on the old records but Derbyshire is well-visited by individuals, groups and colleges and the old records are a good baseline. As might be expected, many of the pre-1980 records were from the Carboniferous and Magnesian limestones and the gritstone in the north of the county. Surprisingly, southern Derbyshire was not well recorded and some species with less than ten records in the north, for example, white bryony (*Bryonia dioica*), were soon eliminated once records from the south accumulated.

The following species, new to the county, were also recorded :-

Killarney Fern	*Trichomanes speciosum*
Marsh Fern	*Thelypteris palustris*
Round-leaved Wintergreen	*Pyrola rotundifolia* spp. *rotundifolia*
Chickweed Wintergreen	*Trientalis europaea*
Cut-leaved Dead-nettle	*Lamium hybridum*
Various-leaved Pondweed	*Potamogeton gramineus*
Sand Sedge	*Carex arenaria*
Blue Moor-grass	*Sesleria caerulea*

From these data a list of rare and threatened higher plants has been compiled for Derbyshire. The major criterion for inclusion, which was decided at the outset, is ten or fewer site records for the plant since January 1980. However, with the diligent surveys in areas of good habitat, for example the Via Gellia, it became necessary to determine whether two or more site records related to the same population. In general sites are regarded as distinct when they are in different 1km grid squares.

National status has been indicated by reference to the Red Data Book (Perring and Farrell, 1983) and *Scarce Plants in Britain* (Stewart et al. 1994). The number under the Red Data Book status (see Perring & Farrell, 1983) is the threat number (TN) which is the result of adding values given to factors like the number of localities (1km squares), an assessment of the likelihood of collecting related to attractiveness, the conservation index relating to protected sites and ease of access. The maximum TN is fifteen although no species has reached it. The Red Data Categories are E = endangered, R = rare, V = vulnerable. In addition, the status 'Scarce' is given for species which, nationally are present in 16-100 ten kilometre squares; these are taken from Stewart et al. (1994).

As the results were analysed three other categories of plants appeared all of which have been excluded from the main list. These are, (with relevant species):-

TABLE 1. Species with no known post-1980 records or where all the post-1980 records are known to be lost.

Western Polypody	*Polypodium cambricum*	
Soft Hornwort	*Ceratophyllum submersum*	
Northern Dock	*Rumex longifolius*	
Tower Mustard	*Arabis glabra*	(Scarce)
Bog-rosemary	*Andromeda polifolia*	
Chaffweed	*Anagallis minima*	
Narrow-leaved Bird's-foot-trefoil	*Lotus glaber*	
Strawberry Clover	*Trifolium fragiferum*	
Shepherd's-needle	*Scandix pecten-veneris*	(Scarce)
Henbane	*Hyoscyamus niger*	
Field Gromwell	*Lithospermum arvense*	
Hound's-tongue	*Cynoglossum officinale*	
Common Calamint	*Clinopodium ascendens*	
Pennyroyal	*Mentha pulegium*	
Green Figwort	*Scrophularia umbrosa*	
Marsh Lousewort	*Pedicularis palustris*	
Yarrow Broomrape	*Orobanche purpurea*	
(British Red Data Book TN=8V)		
Greater Broomrape	*Orobanche rapum-genistae*	(Scarce)

Spiked Rampion *(British Red Data TN=8R)*	*Phyteuma spicatum*	
Meadow Thistle	*Cirsium dissectum*	
Bristly Oxtongue	*Picris echiodes*	
Chamomile	*Chamaemelum nobile*	(Scarce)
Lesser Water-plantain	*Baldellia ranunculoides*	
Floating Water-plantain	*Luronium natans*	(Scarce)
Red Pondweed	*Potamogeton alpinus*	
Blunt-leaved Pondweed	*Potamogeton obtusifolius*	
Opposite-leaved Pondweed	*Groenlandia densa*	
Needle Spike-rush	*Eleocharis acicularis*	
Thin-spiked Wood Sedge	*Carex strigosa*	
Tufted-sedge	*Carex elata*	
Bearded Fescue	*Vulpia ciliata* ssp. *ambigua*	
Meadow Saffron	*Colchicum autumnale*	
Lesser Twayblade	*Listera cordata*	
Small-white Orchid	*Pseudorchis albida*	
Green-winged Orchid	*Orchis morio*	
Man Orchid	*Aceras anthropophorum*	(Scarce)

TABLE 2 Species not included in section B but which just missed inclusion in the main list, i.e. for which there are from 10-19 post-1980 records which are deemed to be wild.

Beech Fern	*Phegopteris connectilis*	
Stinking Hellebore	*Helleborus foetidus*	(Scarce)
Lesser Meadow-rue	*Thalictrum minus*	
Rigid Hornwort	*Ceratophyllum demersum*	
Pale St. John's-Wort	*Hypericum montanum*	
Round-leaved Sundew	*Drosera rotundifolia*	
Bay Willow	*Salix pentandra*	
Wall Whitlowgrass	*Draba muralis*	(Scarce)
Hutchinsia	*Hornungia petraea*	(Scarce)
Yellow Loosestrife	*Lysimachia vulgaris*	
Alternate-leaved Golden-saxifrage	*Chrysosplenium alternifolium*	
Cloudberry	*Rubus chamaemorus*	
Wild Service-tree	*Sorbus torminalis*	
Midland Hawthorn	*Crataegus laevigata*	
Mezereon *(included in 1st Edition of British Red Data Book Vasc. Plants now 'Scarce')*	*Daphne mezereum*	
Hare's-foot Clover	*Trifolium arvense*	
Spurge-laurel	*Daphne laureola*	
Mistletoe	*Viscum album*	
Mare's-tail	*Hippuris vulgaris*	
Clustered Bellflower	*Campanula glomerata*	
Small Teasel	*Dipsacus pilosus*	
Arrowhead	*Sagittaria sagittifolia*	
White Sedge	*Carex curta*	
Bottle Sedge	*Carex rostrata*	

Pendulous Sedge	*Carex pendula*	
Wood Barley	*Hordelymus europaeus*	(Scarce)
Yellow Star-of-Bethlehem	*Gagea lutea*	
Solomon's-seal	*Polygonatum multiflorum*	

TABLE 3. Special species falling into one or more of the following categories:-

(i) with some special Derbyshire connection
(ii) nationally rare but common in Derbyshire
(iii) striking or interesting species

All these species have more than 20 post-1980 records

Moonwort	*Botrychium lunaria*	
Green Spleenwort	*Asplenium trichomanes-ramosum*	
Globeflower	*Trollius europaeus*	
Traveller's-joy	*Clematis vitalba*	
Spring Sandwort	*Minuartia verna*	(Scarce)
Nottingham Catchfly	*Silene nutans*	(Scarce)
Maiden Pink	*Dianthus deltoides*	(Scarce)
Large-leaved Lime	*Tilia platyphyllos*	(Scarce)
Small-leaved Lime	*Tilia cordata*	
Mountain Pansy	*Viola lutea*	
Black Poplar	*Populus nigra* ssp. *betulifolia*	
(The Derbyshire population forms approximately 20% of the national records)		
Narrow-leaved Bitter-cress	*Cardamine impatiens*	(Scarce)
Hoary Whitlowgrass	*Draba incana*	
Pyrenean Scurvygrass	*Cochlearia pyrenaica*	
Alpine Penny-cress	*Thlaspi caerulescens*	(Scarce)
Mountain Currant	*Ribes alpinum*	(Scarce)
Mossy Saxifrage	*Saxifraga hypnoides*	
Grass-of-Parnassus	*Parnassia palustris*	
Spring Cinquefoil	*Potentilla neumanniana*	(Scarce)
(Rock Whitebeam)	*Sorbus rupicola*	(Scarce)
Bloody Crane's-bill	*Geranium sanguineum*	
Yellow-wort	*Blackstonia perfoliata*	
Jacob's-ladder	*Polemonium caeruleum*	
(British Red Data Book TN=6R)		
Melancholy Thistle	*Cirsium heterophyllum*	
Dwarf Thistle	*Cirsium acaule*	
Flowering-rush	*Butomus umbellatus*	
Mountain Melick	*Melica nutans*	
Lily-of-the-Valley	*Convallaria majalis*	
Daffodil	*Narcissus pseudonarcissus* spp. *pseudonarcissus*	

There are few significant, particular threats to higher plants as a group. However as they constitute the majority of the county's vegetation they are subject to the range of general threats to wildlife. Ultimately, these threats are all due to the same cause, more people demanding more from their environment, whether it be directly through requirement for materials, housing, industry, food or leisure, or indirectly through pollution, effluents and the disposal of waste materials. Undoubtedly, past climatic changes have led to the loss of some plants from the county and the restriction in the range of others such as small-leaved lime (*Tilia cordata*).

The account for each species lists the number of post-1980 records, the national Red Data Book status and provides details of the habitat and threats. Order and nomenclature follow Stace (1991) except where no English name is given in the text: in these cases the English name is taken from various sources. The 'Flora' referred to is the Flora of Derbyshire, edited by A.R.Clapham (1969) together with its two supplements (Patrick & Hollick, 1975, Hollick & Patrick, 1980).

Bogbean

		National RDB Status	*No of post-1980 records*

Clubmosses

Fir Clubmoss	*Huperzia selago*		3

Found in small colonies, in central Derbyshire, including one site for rescued plants. Like other clubmosses threatened by man's activities, drought and loss of open habitat.

Stag's-horn Clubmoss	*Lycopodium clavatum*		9

Found in small colonies, from mid-Derbyshire northwards, in sand pits, gritstone quarries or on the shales. Threatened by drought, quarrying and vegetational succession.

Alpine Clubmoss	*Diphasiastrum alpinum*		5

Only five colonies occur, of which all but one are close together. Found mainly in old silica sand pits, in relict heath, where open sand occurs. Threatened by drought, disturbance and vegetational succession.

Horsetails

Rough Horsetail	*Equisetum hyemale*	TN=9E	1

A streamside horsetail for which the one site has been known for over ten years. The main threats are water course management, land-use changes and drying up of the habitat.

		National RDB Status	*No of post-1980 records*

Ferns

Killarney Fern *Trichomanes speciosum* 1

The only site was first recorded in 1991, where it is present only as prothallus. Threats include management, moorland fires and the lack of typical habitat and climatic conditions.

Marsh Fern *Thelypteris palustris* Scarce 1

The only site was first discovered in 1991 on moorland. The main threats are changes in local drainage and moorland fires.

Rustyback *Ceterach officinarum* 5

Found in small colonies, often on man-made structures, e.g, drystone wall, reservoir wall and station platform. Threatened by destruction/clearing of locations.

Oak Fern *Gymnocarpium dryopteris* 10

This woodland fern is rare now probably as a result of the Victorian fern craze. It is often found on rocky ledges in gritstone/shale cloughs which are no longer wooded. (The beech fern, *Phegopteris connectilis*, found in similar places and often with oak fern, just misses inclusion in this list). Grazing may be a threat, as are general habitat changes; the effect of the loss of woodland cover is not known.

Rigid Buckler-fern *Dryopteris submontana* Scarce 2

Two small colonies amongst rocks in limestone dales. Threatened by changes in habitat, following succession.

Soft Shield-fern *Polystichum setiferum* 9

Found mainly in limestone ash woods in the dales, but other sites to the east and south of the Peak District are on other rock types. Some sites have been lost recently, mainly as a result of man's activities.

Conifers

Juniper *Juniperus communis* ssp. *communis* 1

One site near Buxton which has been threatened by site management in the past.

Flowering Plants

White Water-lily *Nymphaea alba* 8

Found in large ponds and lakes, mostly man-made and often with nutrient-poor water. The number of records above does not include known introductions. Losses have occurred due to loss of ponds, either deliberately or through succession and overzealous management.

National RDB Status — *No of post-1980 records*

Green Hellebore *Helleborus viridis* 8

Probably native but may be a garden escape in some sites. Found in woods, hedges and grassland on the limestone. Main threats are changes in grazing and other management.

Greater Spearwort *Ranunculus lingua* 2

Always rare but known to be introduced in other sites; found in small ponds. Main threats are loss of habitat due to deliberate infilling or succession and management. One roadside site particularly vulnerable.

Common Water-crowfoot *Ranunculus aquatilis (s.s.)* 2

Always rare in the county and now only found in ponds in the south. One site is vulnerable to changes in management or pollution from road which has destroyed part of site. The other is in a Trust nature reserve; habitat requirements are poorly understood and care is needed.

Pond Water-crowfoot *Ranunculus peltatus* 2

In old pits. Threatened by infilling and changes in land use.

Fan-leaved Water-crowfoot *Ranunculus circinatus* 8

Majority of records relate to one canal. As with other aquatic species threats are loss of habitat, particularly by succession, pollution and inappropriate management.

Common Meadow-rue *Thalictrum flavum* 1

On canal bank; formerly in wet meadows, stream banks. Main threats are increased provision for recreation and inappropriate management.

Many-seeded Goosefoot *Chenopodium polyspermum* 2

Found in disturbed ground in north-east of county. Sites have been lost to opencast coal extraction, gardening, urban extension and restoration of tips. The two extant sites are threatened by changes in vegetation due to management.

Wood Stitchwort *Stellaria nemorum* 7

Found in damp woods, particularly on the sides of streams. The stronghold is in a Trust reserve but other sites are scattered. The Flora notes the species as decreasing; some sites lost to urban extension but future threats will be woodland management and incursions by stock.

Marsh Stitchwort *Stellaria palustris* 1

Previously recorded for only three sites in the south of the county, the single extant record is for the Peak District. It has been lost from the southern sites due to drainage and agricultural change.

		National RDB Status	*No of post-1980 records*

Field Mouse-ear *Cerastium arvense* 3

Found in dry grassland on the banks of the Trent and on Magnesian limestone. Main threats are erosion and rebuilding of the river bank. The other site is in a Trust nature reserve.

Knotted Pearlwort *Sagina nodosa* 7

Found on sandy soils, spoil heaps and in calcareous turf. Many of the old sites lost to vegetation succession, quarrying and infilling.

Annual Knawel *Scleranthus annuus* 1

Most old records were of gravel and sand workings which have been worked out and restored. The extant record is in a field and is threatened by competition from other plants and agricultural changes.

Night-flowering Catchfly *Silene noctiflora* 1

A weed of sandy soils recorded in north-west Derbyshire but formerly found more widely. The main threat is inappropriate management.

Trailing St. John's-Wort *Hypericum humifusum* 4

Found on dry moorland and sandy soils; two locations in nature reserves but changes in habitat are the greatest threat.

Dwarf Mallow *Malva neglecta* 8

In fields, headlands and verges mostly in the south of Derbyshire. Main threats (and causes of past losses) are changes in management and mineral extraction.

Wild Pansy *Viola tricolor* 2

Found on disturbed ground. One record of *V. tricolor x V. lutea* hybrid with no parent plants recorded. Probably under-recorded although the 'frequent' description in the Flora is not borne out by the records. Main threat is loss of open habitats and cutting before seed set.

Heath Dog-violet *Viola canina* 9

All current records are in north-west Derbyshire on gritstone/shales and heathy areas on limestone. Grazing management will be critical to the survival of this species.

Almond Willow *Salix triandra* 5

All but one of the current sites are in the Trent Valley in damp ground and one is in a nature reserve. Management e.g. indiscriminate clearance is the main threat.

		National RDB Status	*No of post-1980 records*
Creeping Willow	*Salix repens*		1

The historical records cover a range of soil types from limestone to gritstone shales and silica sand. The extant site is on limestone and may be threatened by site management.

Labrador-tea	*Ledum palustre* ssp. *groenlandicum*	TN=6R	3

A rare escape with three sites on the Dark Peak Moors (one not refound) and one not seen since 1981 on Matlock Moor in a plantation. The main threats are moorland fires and recreation pressure, with one site threatened recently by an unofficial footpath.

Shepherd's Cress	*Teesdalia nudicaulis*		1

Found on lead rake; recorded in 1983 but not refound in 1990. Mineral workings and changes in management are the main threats.

Bearberry	*Arctostaphylos uva-ursi*		4

Found around the cloughs, tors and edges of the Derwent Valley on gritstone rocks and banks. The main threats are moorland fires and recreational pressure.

(Intermediate Bilberry)	*Vaccinum x intermedium*		5

Found mainly in the Derwent Valley in the Matlock area and further north on moorland, usually with both parents. Main threats are changes in management and fire which has threatened one site close to Sheffield.

Round-leaved Wintergreen	*Pyrola rotundifolia* ssp. *rotundifolia*	Scarce	1

Not recorded before 1981. Main threat is change in land use/site management

Common Wintergreen	*Pyrola minor*		5

Always rare in the county, occurring in woods and below scrub on railway lines, all except one on Carboniferous limestone. One site disappeared as a result of a tree falling on it, the others are vulnerable to management actions.

Yellow Bird's-nest	*Monotropa hypopitys*		5

All the extant sites are on or close to a disused railway line on Carboniferous limestone and are new since the publication of the Flora. The main threats are habitat changes, both natural and man induced, although all sites are managed by sympathetic owner/manager.

Water-violet	*Hottonia palustris*		6

Several of the traditional sites have been lost due to vigorous management, e.g., long-term drainage of ponds or changes in management, but new sites have been recorded. Some sites are now nature reserves but others may be vulnerable to canal restoration, recreational pressure and agriculture. Availability in garden centres may lead to more deliberate and accidental introductions and possibly to collection. Damage has occurred at one site inadvertently during commercial collection of Daphnia.

		National RDB Status	*No of post-1980 records*
Chickweed Wintergreen	*Trientalis europaea*		1

Found on open moorland, within vice-county 57 but just outside the current administrative county. Main threats are changes in moorland management and fire.

Bog Pimpernel	*Anagallis tenella*		8

Formerly found in suitable habitats in the south of the county but now confined to bogs and damp peaty soil in the north/north-east of the county. Some sites are within SSSIs and are nature reserves. Main threats are habitat changes due to drying out, fire and management.

Navelwort	*Umbilicus rupestris*		2

In addition to the two natural sites known there have been two introductions. Has always been rare and there were only three sites recorded in the Flora. One site has probably been lost due to climbing or other recreational pressure on the gritstone edges.

English Stonecrop	*Sedum anglicum*		4

Found on walls and dry grassy places on gritstone/shales, two within a nature reserve. May be threatened by wall maintenance and other vegetation.

Marsh Cinquefoil	*Potentilla palustris*		9

Originally occurred in a wide range of sites from lowland valley marshes to upland bogs. The major losses have been in the former category as a result of drainage and agricultural improvements. The remaining lowland site is a nature reserve. Upland sites may be threatened by changes in land use although some are protected by sympathetic ownership.

Hoary Cinquefoil	*Potentilla argentea*		3

Found on dry grassy banks on Carboniferous limestone. Only five localities previously recorded. Threats include vegetation succession.

Alpine Cinquefoil	*Potentilla crantzii*	Scarce	1

One site in sparse grassland on limestone in a nature reserve. Another site found recently on a tip, origin unknown. Habitat changes present the most likely threat.

Fragrant Agrimony	*Agrimonia procera*		3

Found in field borders often on Carboniferous limestone, although historical records show a wider range of soils and localities. Causes of loss of ten sites since 1940s unknown, although development and agricultural change are likely.

Burnet Rose	*Rosa pimpinellifolia*		7

Found on Carboniferous limestone, particularly on dalesides where scrub is developing. Several sites within nature reserves; threats include continued scrubbing up and also indiscriminate scrub removal. Possibly under-recorded.

		National RDB Status	*No of post-1980 records*
Sweet-briar	*Rosa rubiginosa*		1

Only two records in the Flora, both with only 10 km square references and neither discovered during this project. One record in urban Chesterfield following introduction.

Wild Liquorice *Astragalus glycophyllos* 7

All the extant records are on Magnesian limestone on colliery tips, banks in quarries, railway banks and verges. Threats include tipping and restoration of the colliery and quarry.

Bird's-Foot *Ornithopus perpusillus* 1

The typical habitat is bare sand or gravel and the remaining site is in a gravel pit nature reserve. The main threat is colonisation as the substrate develops allowing more vigorous plants to grow.

Horseshoe Vetch *Hippocrepis comosa* 3

Occurs on dry limestone outcrops. Main losses appear to have been due to the reduction in sheep grazing and the subsequent changes in habitat.

Wood Vetch *Vicia sylvatica* 4

All of the extant sites are on Carboniferous limestone in shade. The main threats are changes in the habitat and scrub clearance.

Smooth Tare *Vicia tetrasperma* 2

In grassy disturbed places, where the main threats are land use changes including site restoration.

Narrow-leaved Everlasting-pea *Lathyrus sylvestris* 1

The four historical records are on canal towpaths, railway banks or tips. The one remaining site is within a nature reserve where vegetation succession is a threat.

Spiny Restharrow *Ononis spinosa* 9

Found in disturbed grassy places and banks on a range of soils including floodbanks beside the Trent. Twelve pre-1980 sites lost to development and changes in land use.

Knotted Clover *Trifolium striatum* 7

Found in short grassland/open habitats on a variety of soils. Sites have been lost as a result of habitat change, which is the main threat to most of the remaining sites.

Subterranean Clover *Trifolium subterraneum* 2

The three recorded sites are all in the extreme south-east of the county near to the River Trent on sandy gravelly soils. Threats to the remaining sites are scouring by flood water and flood protection works.

		National RDB Status	*No of post-1980 records*

Petty Whin *Genista anglica* 3

Found on moorland in better soils, usually where unaffected by burning, mostly in the Derwent Valley although one site is in a lowland valley. The main threats are changes in management.

Whorled Water-milfoil *Myriophyllum verticillatum* Scarce 1

Only two sites identified and the one which was extant in the early 1980s may have been lost with the drying out of its habitat because of lack of management and alterations to the drainage due to opencast workings.

Alternate Water-milfoil *Myriophyllum alterniflorum* 2

No recent records noted in the Flora but three sites have been found since. The two current pond sites are both under pressure from development, and pollution or disturbance are the main threats.

Water-purslane *Lythrum portula* 3

Found in bare ground on the edge of ponds on neutral or acid substrate; two sites being beside reservoirs. Main threats are lengthy drawdown and inappropriate management.

Upland Enchanter's-nightshade *Circaea* x *intermedia* 2

In spite of the name the majority of the historical records were in lowland Derbyshire in woods and shady places. One of the extant records is in an SSSI, the other in a nature reserve, but since the species requirements are unknown, precise threats are unidentifiable.

Dwarf Spurge *Euphorbia exigua* 4

Found on cultivated and waste ground in the north-east and north-west of the county. The Flora notes that there are no recent records from the north. Vulnerable to development, changes in agricultural practices and vegetation.

Alder Buckthorn *Frangula alnus* 3

Recorded from a range of damp soils, on Magnesian limestone, alluvium and gritstone/shales and formerly on Carboniferous limestone; seven sites have been lost. All the extant sites either protected as a nature reserve or under special management.

Wood Crane's-bill *Geranium sylvaticum* 1

This site found in 1980 was the first record for fifty years, but by 1993 the plants had disappeared, although seeds have been taken and a cultivated population exists. The cause of the loss is unknown.

		National RDB Status	*No of post-1980 records*
Long-stalked Crane's-bill	*Geranium columbinum*		8

Found on grassy banks in shallow soil on limestone, including quarries. Main losses of sites due to changes in management resulting in habitat progression.

Hedgerow Crane's-bill	*Geranium pyrenaicum*		5

Found in hedgerows and disturbed ground; sixteen pre-1980 records not refound. Main threats are inappropriate management and succession.

Bur Parsley	*Anthriscus caucalis*		1

Of the four Flora sites only one is now extant on Magnesian limestone and in a nature reserve. A plant of open ground: the main threat is change in the vegetation.

Buck's-horn Plantain	*Plantago coronopus*		1

None of the three Flora sites were refound but in 1991 a new site was found by a reservoir in South Derbyshire. The main threat will be loss of suitable habitat or destruction of the site by recreational pressure.

Tubular Water-dropwort	*Oenanthe fistulosa*		8

Found in ponds, ditches and marshy places in the south and south-east of the county. Two sites are in nature reserves. The main threats are inappropriate management, tipping and major dredging.

Narrow-leaved Water-dropwort	*Oenanthe silaifolia*	Scarce	1

This record is in an atypical habitat in a damp area on a north-facing limestone daleside which is a nature reserve. The main threats are inadvertent grazing or picking.

Hemlock Water-dropwort	*Oenanthe crocata*		2

Not refound in Flora location, but two new sites were found in an oxbow of the Trent and the Erewash Canal. Both are threatened by management work and the current use.

Fine-leaved Water-dropwort	*Oenanthe aquatica*		5

Eight historical records have been confined to pools and marshes close to the River Trent. Although this species can survive the habitat drying up, the main threats are successional change towards carr or infilling.

Pepper-saxifrage	*Silaum silaus*		4

This species which prefers undisturbed grassland, whilst only occurring occasionally in the county, has declined recently as a result of the agricultural improvement of meadows. None of the sites is protected and they are vulnerable to agricultural change.

		National RDB Status	*No of post-1980 records*
Lesser Marshwort	*Apium inundatum*		1

Only ten historical records exist for this plant of ponds, lakes, canals and ditches; the one extant site is on a canal feeder reservoir. It has been lost from canals due to neglect and disuse but the cause of loss from lakes and some ponds is unknown.

Stone Parsley *Sison amomum* 1

Normally found on hedgebanks; the only site in the Flora has not been refound, but the species has been recorded in a garden in Chesterfield. Change of management/ownership is the main threat.

Knotted Hedge-parsley *Torilis nodosa* 1

Always a rare plant in the county (only two sites in the Flora); found on dry banks, the only extant site being a new one. The main threat must be vegetation or land use change.

Field Gentian *Gentianella campestris* 1

Past records were from pastures, especially on non-calcareous soils. i.e., the Millstone Grits/Edale Shale Series and the current site is on this rock type. Land use changes account for some of the lost eight sites; the current site may be threatened by management or a lack of it.

Deadly Nightshade *Atropa belladonna* 2

This species has always been rare with only four known sites. One of the two sites seen in the 1980s was not refound in 1990 but this plant of woods, scrub and rough ground will suffer from cultivation. The only extant site is in a garden.

Bogbean *Menyanthes trifoliata* 8

Although concentrated on the eastern gritstone moors, another northern site has been identified recently and one southern valley site remains. Introductions into three SSSI nature reserves are not included in the total. The main threats are collection and habitat change.

Creeping Forget-me-not *Myosotis secunda* 1

Although only one site has been re-visited post 1980 it is likely that it still occurs in upland flushes on the gritstone in the Dark Peak and on the Eastern Moors in previously recorded sites. Threats include changes in land management although in many sites these are now unlikely, unless tree planting or other changes are undertaken within the North Peak ESA area.

Field Woundwort *Stachys arvensis* 5

Found in disturbed, non-calcareous soils in sites from south to north-east Derbyshire; garden sites not included. Main threats are changes in land use or intensification of recreational use on waste ground and habitat progression.

		National RDB Status	*No of post-1980 records*
Cut-leaved Dead-nettle	*Lamium hybridum*		1

Discovered on a field edge in the east of the county during the period of survey for this book. Main threats are land use change or herbicide.

Common Gromwell *Lithospermum officinale* 6

Found in woods and shady places on both Carboniferous and Magnesian limestones, with a good percentage of all sites refound. Several sites are within SSSIs. The main threats are drastic woodland/scrub management and quarry restoration.

Vervain *Verbena officinalis* 1

None of the three historical sites have been refound but a new site in south-east Derbyshire was found on an old tip, where habitat succession is the main threat, even though it is a nature reserve.

Red Hemp-nettle *Galeopsis angustifolia* Scarce 5

Found on thin grassland, waste ground and in cultivated fields, mostly on Carboniferous limestone but with one site near Chesterfield. Main threats are agricultural management, both existing and changes.

Lesser Skullcap *Scutellaria minor* 3

Always rare, found in wet moorland and open woodland on gritstone and shales in the north of the county. Main threats are changes in moorland management, especially drainage and recreation pressure.

Cat-mint *Nepeta cataria* 1

Always rare, with only four historical sites; the current site is on Magnesian limestone. Found in open grassland, where the main threat is habitat change.

Shoreweed *Littorella uniflora* 6

All but one of the Flora sites have been refound, although two of them are under threat from development (which was opposed). The typical habitat of shallow lake margins or exposed shore is vulnerable to long-term drawdown or other unsympathetic management.

Dark Mullein *Verbascum nigrum* 6

Found on waste ground and banks, almost exclusively on Carboniferous limestone. The main threats are habitat change and management.

Mudwort *Limosella aquatica* Scarce 2

The Flora recorded two sites, one of which had been destroyed, the other being still extant. The species survives on bare mud at the edge of the reservoir which is a canal feeder. The site is an SSSI but the main threat is inappropriate management of the water levels due to the need to maintain the canal system. Recently recorded from Derwent Reservoirs on ground exposed by drawdown during drought.

		National RDB Status	No of post-1980 records
Sharp-leaved Fluellen	*Kickxia elatine*		1

Found on field borders and in disturbed ground beside paths. Main threat is management including herbicide use.

Pink Water-speedwell *Veronica catenata* 10

This species was not recorded in the Flora; it is found in shallow water, mostly in the Trent Valley but also in the north-east on Magnesian limestone. Some of the current sites will be affected by gravel pit restoration, others by management and some by natural habitat change. Three sites lie within nature reserves, two of which are SSSIs.

Common Broomrape *Orobanche minor* 2

Only four sites previously recorded, one not seen since 1942 and one in a garden where it may have been introduced. The two extant sites are on disused railway lines, one of which has planning permission for a major development. The Trust is involved in plans to retain the main part of the colony, although changes in use of the site by local people may be a future threat.

Ivy-leaved Bellflower *Wahlenbergia hederacea* 4

Found in damp acid soils, particularly in cloughs in north Derbyshire. Of a total of seventeen past records, ten are from 1979 onwards, compared with only two sites in the Flora, of which one has not been refound. The main threats are changes in management, possibly funded with incentives.

Sheep's-bit *Jasione montana* 5

Found on dry grassy banks away from the limestone. Some sites have been lost to development including road alterations. Others were not refound although there was suitable habitat and in other cases habitat change had occurred.

Mountain Everlasting *Antennaria dioica* 4

In Derbyshire, usually found on limestone, in short grassland. The main threats are changes in habitat.

Heath Cudweed *Gnaphalium sylvaticum* 2

Found in dry open woods and grassland on acid soils. Of five sites in the Flora two were thought to be extinct in 1969. Four more sites have been found since then but only two remain. The site threats include changes of use and management.

Cornflower *Centaurea cyanus* Scarce 3

Found on cultivated and waste ground but by 1969 was 'very uncommon'. One known introduced site not included. Main threats are changes in land use or management.

National RDB Status | *No of post-1980 records*

Dwarf Elder *Sambucus ebulus* 5

This species which is described as possibly native is found on roadsides and damp grassy places, possibly as a result of past cultivation. Two sites have been lost, one of them caused by overshading. This and inappropriate verge management, are the two main threats to the species.

Hedge Bedstraw *Galium mollugo* ssp. *erectum* 3

A rare plant of dry roadsides and hedgebanks with only three known sites. The sites vary in their susceptibility to habitat changes and other pressures, two being on verges and one at a site with considerable visitor pressure.

Woolly Thistle *Cirsium eriophorum* 7

Found in dry grassland mostly on limestone, always uncommon but decreasing due to changes in habitat, mostly as a result of management but also as a result of reclamation following reworking of mineral sites.

Greater Burdock *Arctium lappa* 2

Always very rare in the county, found in waste places; both sites were recorded in the south of the county. Main threat is change in land use.

Angular Solomon's Seal

Narrow-leaved Water-plantain *Alisma lanceolatum* 3

The sites are within an SSSI, managed as a nature reserve. Habitat change due to outside circumstances, e.g. drying out of the shallow water habitat, pollution, etc. are main threats. Although not listed in the British Red Data Book, the IUCN List notes it as V (vulnerable).

		National RDB Status	*No of post-1980 records*

Shining Pondweed *Potamogeton lucens* 1

Always rare but formerly all sites were in the south of the county; the extant site is in the north, in an industrial site. The main threat is reclamation or inappropriate management of the pond where it occurs.

Various-leaved Pondweed *Potamogeton gramineus* 1

First record in mid-1980s in an SSSI, where the main threat is vegetation change.

Long-stalked Pondweed *Potamogeton praelongus* 2

Always very rare, now recorded in a borrow pit and the River Trent, formerly found in one other site in the Trent Valley. Management, flooding and restoration of the borrow pit are the main threats.

Flat-stalked Pondweed *Potamogeton friesii* 1

Found in canals and slow streams, but always rare and lost from sites such as the Derby Canal due to infilling and the Trent and Mersey Canal possibly due to increased recreational use. The current site is threatened by restoration.

Lesser Pondweed *Potamogeton pusillus* 2

Always rare; both sites are within feet of the county boundary, one in the south west and the other in the north-east, in a lake and a stream. Both lie within sites where nature conservation is taken into account.

Small Pondweed *Potamogeton berchtoldii* 4

There was confusion in the past between this species and the preceding one but both were always rare. Found in lakes and small ponds and threatened by inappropriate/ inadvertent management.

Grass-wrack Pondweed *Potamogeton compressus* Scarce 1

Always rare, found chiefly in canals and canalised sections of rivers; the site is threatened by canal restoration.

Horned Pondweed *Zannichellia palustris* 3

Always very uncommon; the main habitats are lakes and a gravel pit. The main threats are inappropriate management or habitat changes, e.g. succession.

Round-fruited Rush *Juncus compressus* 3

Always very rare; found in damp habitats on an industrial site, railway sidings and within a wood. Threats due to re-development and recreation.

Blunt-flowered Rush *Juncus subnodulosus* 1

Always very rare, found in a lowland marsh, although all other past records have been on Magnesian limestone. Main threat is vegetation succession.

	National RDB Status	No of post-1980 records

Broad-leaved Cottongrass *Eriophorum latifolium* 2

Two different sites occur within the same wood on Magnesian limestone; the species has always been very rare, only two other sites having been recorded in the past. The main threats are drying out of the wet habitat and succession.

Grey Club-rush *Schoenoplectus tabernaemontani* 1

Always very rare but not refound in two Flora localities. Found in marsh in gravel pit nature reserve in the Trent Valley. Main threat is habitat change.

Many-stalked Spike-rush *Eleocharis multicaulis* 2

Not recorded in the Flora. Two sites from bog areas on moorland on the east side of the Peak District. Main threat due to management changes.

Few-flowered Spike-rush *Eleocharis quinqueflora* 1

Still found in the Flora site on Magnesian limestone. Also a record in 1970s for an earlier site. Main threats are drying out of damp area.

Flat Sedge *Blysmus compressus* 5

Found in marshy areas on Carboniferous limestone, several in SSSIs and nature reserves. Main threats are drainage of land and habitat change.

CAREX SPECIES

Although efforts were made to check all relevant sedge records, the numbers of sites in the following accounts may be underestimates due to the difficulties of finding sedges at the correct stages for identification.

Prickly Sedge *Carex muricata* ssp. *lamprocarpa* 1

Found on dry grassy banks in the south of the county. The main threat is a change in land use although there is also a planned road scheme.

Grey Sedge *Carex divulsa* ssp. *leersii* 5

Found in open woods, banks and waste ground usually on limestone. Only two sites noted in the Flora, but a total of ten sites have been recorded now, although several have been lost in the 1980s due to quarrying. The main threats to the remaining sites are vegetational change; most of the sites are within SSSIs.

Sand Sedge *Carex arenaria* 1

Found on a disused railway sidings in the cinder area. The main threats are proposed developments and the reinstatement of the railway line, also alterations to the drainage as well as vegetational succession.

		National RDB Status	No of post-1980 records

Brown Sedge *Carex disticha* 4

Found in marshy stream margins and wet grassland over a wide area but still not common, only fourteen sites known (five from the Flora). Main threats are drainage and land use changes.

Dioecious Sedge *Carex dioica* 1

Very rare; only four sites in the Flora, one of which is still extant in base rich flushes on Magnesian limestone. One of the other sites has dried up and changed and this must be the main threat for the remaining site although it is in a site where management for conservation occurs.

Cyperus Sedge *Carex pseudocyperus* 4

Found by the side of streams in permanently wet ground all in South Derbyshire. Rare, only five sites listed in the Flora and ten site records previously recorded in total. The main threats are water levels and inappropriate management.

Bladder Sedge *Carex vesicaria* 5

Always uncommon but found historically in most parts of the county; now restricted to the north and east of the county. The main threats are inappropriate management of the ponds and streamsides.

Tawny Sedge *Carex hostiana* 1

Only four sites noted in the Flora although a total of nine now listed, of which the only one recorded since 1980 is on Magnesian limestone in a site managed with nature conservation in mind. Main threats are to the maintenance of the flush habitat, e.g. drainage and succession.

Long-stalked Yellow-sedge *Carex viridula* ssp. *brachyrrhyncha* 4

Recorded in the Flora as occasional although only four sites are listed. Past records from nine sites now known. Three of the extant sites are within nature reserves and one site within a wood where nature conservation is taken into account. The main threats are to the flush habitats, management and changes in water supply.

Small-fruited Yellow-sedge *Carex viridula* ssp. *viridula* 1

Only one site ever known in a damp area on Magnesian limestone in a site where nature conservation is taken into consideration. The main threat is change in water supply/ drainage.

Pale Sedge *Carex pallescens* 6

In damp, open woodland and in marshy grassland, only rarely on limestone. Seven sites recorded in the Flora but twenty two past records now known. All current sites in north Derbyshire since the southern sites were not refound. Main threats are changes in land use and reclamation.

use and reclamation.

		National RDB Status	*No of post-1980 records*
Fingered Sedge	*Carex digitata*	Scarce	8

Found in short open turf on Carboniferous limestone on rocky slopes and ledges. Only three sites in the Flora. Main threats are vegetation change although several sites are within SSSIs and/or nature reserves.

Bird's-foot Sedge	*Carex ornithopoda*	TN=3R	5

Always rare but at each of the two Flora locations there are six sub-localities all very close together; three new sites have been found. All sites are within a SSSI and some are within nature reserves. The main threats are vegetation changes leading to competition.

Rare Spring-sedge	*Carex ericetorum*	Scarce	2

Always very rare, with only one site in the Flora, in short turf on dry shallow soils on Magnesian limestone. Both extant sites are threatened by changes in management although one is within an SSSI.

Soft-leaved Sedge	*Carex montana*	Scarce	4

Always very rare with only one site in the Flora; all four extant sites are very close together but in separate locations, on rough grassy banks and under shade on Magnesian limestone. All threatened by vegetational change even though three are within an SSSI.

Wood Fescue	*Festuca altissima*		4

Always very rare; only one site in Flora, now increased by three. Found on rocky slopes under woodland on limestone and gritstone. All within SSSIs. Main threats inadvertent management, recreational pressure and woodland felling.

Squirreltail Fescue	*Vulpia bromoides*		10

Only six sites in the Flora, with twenty six recorded in the past, but only ten current sites. Found on dry grassy places. The main threats are vegetation succession and re-development.

Rat's-tail Fescue	*Vulpia myuros*		7

No sites recorded in the Flora. All sites in south east of the county on derelict sites, e.g. railway sidings, rifle range and warehouse, rough ground. Threats are re-development, canal restoration and habitat succession.

Reflexed Saltmarsh-grass	*Puccinellia distans*		5

Found on roadside verges on moorland, where it is maintained by winter road salting. Generally increasing and greatly under-recorded.

Spreading Meadow-grass	*Poa humilis*		4

Recorded for only one site in the Flora but noted as often overlooked; probably a widespread component of upland pastures. Found on moorland verges and wood edges.

		National RDB Status	*No of post-1980 records*
Whorl-grass	*Catabrosa aquatica*		4

Always rare; only four sites in the Flora and a total of nine previously recorded, three of the four extant sites are in the same brook in south-west Derbyshire and the other in a pond in the south-east. The main threats are management change and pollution.

Blue Moor-grass	*Sesleria caerulea*	Scarce	1

Not recorded in the Flora. Found in sparse grassland, on limestone. Main threat is land use change affecting habitat composition; recorded in nature reserve.

Purple Small-reed	*Calamagrostis canescens*		2

Not recorded in Flora; found in wet areas in woodland on Magnesian limestone. Main threat is habitat change although site is managed for nature conservation.

Wood Small-reed	*Calamagrostis epigejos*		10

Found in damp woods, scrub, on disused railway sidings, car parks, along paths; only three sites recorded in the Flora but a total of seventeen old sites now known. Main threats are inappropriate management and re-development.

Orange Foxtail	*Alopecurus aequalis*		2

Found in wet grassland in north-east of the county, rare. Main threats are management and land use changes.

Smooth Brome	*Bromus racemosus*		3

In grassy places; only six sites in the Flora, although twenty six old sites now recorded; only three sites recorded in the 1980s. Main threat is agricultural change.

Meadow Barley	*Hordeum secalinum*		3

Found in meadows, only seven sites recorded in the Flora, although twenty old sites now recorded. The three extant sites are on the edge of urban areas where recreational pressures and land use changes are the main threats.

Branched Bur-reed	*Sparganium erectum* ssp. *neglectum*		2

By ponds, ditches and canals, always very rare but restricted to ponds now, including one SSSI managed as a nature reserve. Main threat is habitat change and inappropriate management on the other site from which it may have been lost in the late 1980s as a result of clearance works.

Lesser Bulrush	*Typha angustifolia*		10

Found in ponds, canals and slow-flowing brooks; always uncommon and two sites found within the 1980s lost to development and management. One ornamental lake has five separate colonies but it has been lost from other similar lakes. Main threat is

		National RDB Status	*No of post-1980 records*
Angular Solomon's-seal	*Polygonatum odoratum*		2

Always very rare and found on the ledges of limestone cliffs often in shade. Main threats are habitat changes and recreational pressure.

Star-of-Bethlehem	*Ornithogalum angustifolium*		5

In grassy places, rare, but the Flora describes it as 'probably introduced'. Found on Carboniferous limestone. Main threats are changes in habitat.

Field Garlic	*Allium oleraceum*		8

Found on ledges and grassy places in Carboniferous limestone dales and on Magnesian limestone. The main threats are changes in management and land use.

Sand Leek	*Allium scorodoprasum*		3

Always very rare, found in dry grassland in Wye Valley. Main threats are changes in agricultural land use.

Marsh Helleborine	*Epipactis palustris*		1

Only one current site listed in the Flora, although reported to be lost from several former localities. The same site remains now within an SSSI although water abstraction/drainage has triggered harmful habitat changes in the marsh habitat.

Dark-red Helleborine	*Epipactis atrorubens*	Scarce	4

Found on limestone rocks and screes, usually in woodland or scrub. The four extant sites are within nature reserves and/or SSSIs. The main threats are habitat changes, e.g. due to grazing etc.

Green-flowered Helleborine	*Epipactis phyllanthes*	Scarce	1

The only historical site still remains; the site is managed with nature conservation in mind but changes in drainage, e.g. by mine pumping and other local works, may be threats to the damp woodland habitat.

Bird's-nest Orchid	*Neottia nidus-avis*		6

Found in shady places on Carboniferous and Magnesian limestone. Always very rare, only two sites listed in Flora but nine old site records. Main threats are forestry management and recreational pressure.

Pyramidal Orchid	*Anacamptis pyramidalis*		10

Found in limestone grassland on banks, always uncommon. Some sites within SSSIs or in nature reserves, but others threatened and destroyed by quarry extensions and workings and others by lack of grazing.

		National RDB Status	*No of post-1980 records*
Early Marsh-orchid	*Dactylorhiza incarnata*		1

The Flora reports no recent records but in 1989 it was found in an SSSI managed as a nature reserve in the south of the county. The main threats are drying out of the marshy habitat and succession.

Northern Marsh-orchid *Dactylorhiza purpurella* 8

Found in marshy ground in north-west Derbyshire, usually in separate locations from the Southern Marsh-orchid (*D. praetermissa*). Several sites in quarries or on old workings and therefore threatened by re-working and restoration as well as succession.

Burnt Orchid *Orchis ustulata* Scarce 10+

This species is found in a very well defined small area of the county in short grassland on Carboniferous limestone. There are more than ten records but many are very close and may refer to the same site. Main threat is inappropriate management. All sites in private ownership, unavailable for purchase.

Fly Orchid *Ophrys insectifera* 7

Found in open limestone woods, scrub, grassland, quarry floors and spoilheaps. Always rare. Some sites within SSSIs, others not protected. Main threats are habitat change, mineral extraction and site restoration.

Pyramidal orchid

ACKNOWLEDGEMENTS

Many botanists have contributed to the records, either as individuals or as part of group efforts, e.g. Matlock Field Club and Lyme Natural History Society. The Sorby Natural History Society allowed us access to their records.

Some devoted their time to special groups, e.g. the clubmosses and black poplars and others ranged far and wide tracking down as many sites and species as possible. The following checked 10 km squares or provided significant numbers of records.

Ken Allsop	Penny Anderson	George Challenger
Mike Coveney	Dr. Trevor Elkington	Roy Frost
Martin Grace	Margaret Hewitt	Jim Russell
Robin Slaney	Roy Smith	Dr. Lawrence Storer
Trevor Taylor	Rhodri Thomas	Eileen Thorpe
Marion Torry	Grace & George Wheeldon	Dr. Alan Willmot

REFERENCES

CLAPHAM, A. R. ed. (1969) *Flora of Derbyshire.* County Borough of Derby Museum & Art Gallery, Derby

HOLLICK, K. M. & PATRICK, S. (1980) *Supplement to Flora of Derbyshire 1969. Additional Records Received* 1974-1979. Derby City Council, Museums and Art Gallery, Derby

PATRICK, S. & HOLLICK, K.M. (1975) *Supplement to Flora of Derbyshire 1969. Additional Records* 1969-1974. Derby Borough Council, Museums & Art Gallery, Derby

PERRING, F. H. & FARRELL, L. (1983) *British Red Data Books: 1 Vascular Plants.* 2nd Edition. Royal Society for Nature Conservation. Lincoln

STACE, C. (1991) *New Flora of the British Isles.* Cambridge University Press, Cambridge

STEWART, A., PEARMAN, D. A. & PRESTON, C. D. eds. (1994) *Scarce Plants in Britain.* Joint Nature Conservation Committee, Peterborough

INDEX
English Names

Latin
Scientific Names (Genera)

MOLLUSCS

PETER TATTERSFIELD

The phylum Mollusca is the second largest in the animal kingdom. It contains a wide variety of animals, ranging from the marine octopus, sea slug, winkle and cockle, to the familiar garden snail and slug. British non-marine molluscs fall into two classes, the gastropods which include terrestrial and freshwater snails and slugs, and the exclusively aquatic bivalves or freshwater mussels or pea shells.

With the possible exception of the large tracts of blanket bog in the north of the county, snails and slugs occur in all the habitat types in Derbyshire. Old, deciduous woodlands typically have the richest land mollusc faunas but wetlands, grasslands and even flushes and seepages on unenclosed moorland or in acidic grassland can all support distinctive suites of species. The fact that some species are generally confined to particular conditions means that they can be useful as habitat 'indicators' (Kerney & Stubbs 1980). For example, some are confined to exceptionally well-established habitats such as ancient woodland or species-rich calcareous grassland whereas others are found in places which have been disturbed by man's activities. Amongst freshwater species, some need clean, flowing water whereas others are tolerant of polluted conditions and some are generally associated with water bodies which dry up in the summer.

Mollusc recording in Derbyshire

Mollusc recording in Derbyshire was probably in its heyday in the latter part of the 19th century, when L. E. Adams, H. Milnes, T. Hey, J. W. Jackson and others contributed records to the national non-marine mollusc census and compiled lists for the county. There was a lull in recording activity during the first 60-70 years of the 20th century but a resurgence of interest occurred more recently when distributional data was being collected for the National Atlas (Kerney 1976). Since then, additional species have been added to the vice-county (v.c. 57) list which currently stands at 123 (Kerney 1982), comprising 26 freshwater snails, 19 bivalves and 78 terrestrial or marsh snails or slugs. To put this figure in context, the county list represents about 62% of the total number of species of non-marine mollusc known from the British Isles, which currently stands at 197.

The selection of the Red Data Book mollusc species

The chequered history of the interest in Derbyshire's snails and slugs and the uneven intensity of recording has implications for the selection of the Red Data Book species. For example, most of the old records have poor and imprecise locational information and very few include any information about population size, habitat type or potential threats. However, since most of the old records were scrupulously verified for inclusion in the national vice-county census, and at least some are supported by specimens in museums, they have been included in the selection procedure. A further important reason for including the old records is that several of the species have not been recorded in the county since, and it would not be appropriate to exclude them from consideration solely on the grounds of poor quality locational data.

The RDB mollusc species have been selected from those for which there have been ten or fewer records from the county. All records have been considered, irrespective of date, although some particularly poorly defined records have been omitted. In addition to the number and distribution of Derbyshire records, the following factors have also been considered during the selection of the RDB species: (1) whether the species is restricted to a declining habitat such as unimproved grassland or ancient woodland; (2) whether the species is on the edge of its current British range in Derbyshire; (3) its sensitivity to pollution or disturbance; (4) its national distribution, and especially whether the species has declined in the recent past; (5) whether the species has suffered from deleterious and widespread changes in habitat management.

Each of the twenty six RDB species has been placed in one of three categories (see Table 1):

Rare species are typically known from a very small number of sites (usually one or two). In addition, most of these species have habitat requirements and national distribution patterns which suggest that they are undoubtedly uncommon and probably declining in the county and country.

Local species are more frequent than the rare species, but they are either highly localised or have a very thinly scattered distribution in the county.

Species *not seen recently* were typically last recorded in the 19th or early 20th century. A provisional status has been assigned to some species and the reasons for this are described in the relevant species account.

Helicella itala

TABLE 1. Numbers of species in each of the Red Data Book categories

Category	Number of species		
	Terrestrial	Freshwater	Total
Rare	6	3	9
Local	9	4	13
Not seen recently	2	2	4
Total	17	9	26

Species which have been excluded

Some species which are known from less than ten sites in the county have not been included in the RDB for one or more of the following reasons: (1) their national status and distribution suggests that they are probably more common in the county than indicated by current records; (2) they have only recently been introduced into the county; (3) they are critical taxa which are difficult to identify and are only recently being identified by field workers. This applies especially to some slugs in the genus Arion; (4) although reported from the county, they have not been confirmed for the national vice-county census and their presence must therefore remain uncertain.

These excluded species are listed at the end of the species accounts.

In the following accounts, separate lists are given for freshwater and terrestrial species which are presented in systematic order. Each account gives some idea of the species' national distribution, its status as assigned by JNCC, habitat requirements, the Derbyshire records and distribution and, where possible, some indication of the most likely threats it faces. In addition, the number of sites which are protected, for example, as Derbyshire Wildlife Trust reserves, SSSIs or NNRs are also noted, where these can be identified.

Nomenclature and arrangement follows Kerney (1976). National status is based on the Invertebrate Site Register (JNCC) categories.

National Status

The national status follows the classification in the Invertebrate Site Register (JNCC). Relevant categories are defined as follows:

Nb — Nationally notable (Scarce) Category B

Taxa which do not fall within national RDB categories but which are nonetheless uncommon in Great Britain and thought to occur in between 31 and 100 10km squares of the National Grid.

Local

This is a less formal category used by the ISR to cover species which are not common. They may be widespread but restricted to vulnerable habitats. Many are useful indicators of a particular habitat.

TERRESTRIAL SPECIES

	National Status	*Derbys. status*
Acicula fusca	Nb	Rare

This small, operculate species was first found in the county in 1983 in a woodland SSSI, also a Derbyshire Wildlife Trust reserve. It has since been located at only one other site; both are in the east of the county. *Acicula* has a widely scattered but very local distribution in Britain. It is found amongst leaf litter or moss in old, damp woodland. Threats include loss of ancient woodland and disturbance and compression of fragile, wet soils and leaf litter beds.

	National Status	*Derbys. status*
Columella aspera	Local	Local

Known from two ancient woodland sites in the county, both SSSIs. It is a local species in Britain, typically found in upland areas where it is often associated with woodland, marshes and even bracken beds. It is surprising that it is not more widespread in the Peak District, although sufficient searching has been undertaken to indicate that it is truly uncommon.

Vertigo pusilla	Nb	Rare

Currently known from a derelict wall in one limestone dale (SSSI & NNR), but previously recorded from Matlock and Dovedale in the latter part of the 19th century. A very local species in Britain, often associated with dry stone walls, rock habitats or screes.

Vertigo antivertigo	Local	Rare

First recorded in Derbyshire from the marshy edges of the Cromford Canal (SSSI) in 1983, but now also known from a sedge-rich flush in the Dark Peak. Threats include drainage of wetlands.

Abida secale	Nb	Not seen recently

Recorded from Miller's Dale, Monsal Dale and from near Haddon Hall at around the turn of the 18th century but not seen since despite a search in suitable places. A local species in Britain which is associated with calcareous grasslands and limestone rock habitats. Declining as a result of loss of grasslands and cessation or relaxation of grazing regimes.

Leiostyla anglica	Local	Local

About eight sites (including three SSSI's and one Derbyshire Wildlife Trust reserve) are known for this species, all in the Dark Peak. It is usually associated with particularly well-established habitats, especially either floristically rich flushes or ancient, wet woodland. Such sites are declining and of great conservation significance. Although local in Britain, *Leiostyla* is commonest in the north and west of the country and Derbyshire is close to the eastern edge of its main range thus increasing the significance of the county's populations. Threats include drainage of wetlands, eutrophication through fertiliser application or run-off and loss of old woodland.

Vallonia pulchella		Rare p

One record from the lowland part of the county. Not uncommon in England generally, usually found in calcareous marshes or damp grassland. The scarcity of these habitats in the county may account for its apparent rarity. Although it is clearly scarce in the county, it has been assigned a provisional Rare status because of the possibility that it has been under-recorded.

	National Status	*Derbys. status*

Spermodea lamellata — Local — Not seen recently

Recorded from Hall Dale Wood in Darley Dale and Monsal Dale in 1890 but not since seen despite recent searches at these sites. A local species in Britain, usually associated with old deciduous woodland in north and west Britain.

Milax gagates — Local — Rare

Two records; this slug was first recorded from a cemetery wall in 1982 and subsequently found in the Dovedale area. Scattered in Britain but commonest on the coast in the west; often associated with man elsewhere.

Limax tenellus — Nb — Rare

Known from two ancient woodland sites (both SSSIs). A very local but widely scattered slug in Britain. It is most readily found in the autumn when it comes to maturity and can be found feeding on fungi. Threats include the loss of ancient woodland.

Testacella scutulum — Local — Rare

First recorded in 1897 from Little Eaton and since found on a roadside at Bubnell. This slug species is usually found in parks and gardens or other places used by man. It carries a small external shell at the hind end and is carnivorous, preying on earthworms. The partly subterranean habit of this species makes its presence difficult to detect.

Candidula intersecta — Local

Two late 19th century records, but recently recorded only from four old limestone quarries and on a road verge. This lime demanding species is common in the south of England; its British distribution extends northwards along the coast.

Cernuella virgata — Local

Three late 19th century records from east Derbyshire but otherwise known only from the Tideswell area (several recent records), and a small area of short grassland in Winnats Pass (1992). A calcareous grassland species which, in Britain, is common south and east of a line from the Severn to the Humber and on coastal sand dunes. The Derbyshire colonies are on the north-eastern edge of the species' inland range which increase their significance.

Helicella itala — Local — Local

Recently recorded from seven relatively discrete limestone areas including several SSSI's and a Derbyshire Wildlife Trust reserve. Also known from three pre-1900 records. A strict calcicole usually found on south facing slopes in high quality calcareous grassland where it prefers short, well-grazed grassland. It is a local and declining species in Britain and is threatened by loss of calcareous grassland, agricultural improvement and changes in grazing regimes.

	National Status	*Derbys. status*
Monacha cantiana	Local	Local

Known from about four relatively discrete areas of the county (with several independent records from some areas). In Derbyshire, this species occurs especially on road verges and one of the sites is on a rubbish tip. Very common in the south and east of England but on the edge of its range in Derbyshire. However, there is some evidence that this species may be extending its range north-westwards, so it may become more frequent in Derbyshire in the future.

Ashfordia granulata	Local	Local

Apparently a very localised species in the county, being recorded from only four recent sites in the area between Castleton and Calver. In Britain, this species is usually associated with damp and shaded places such as wet woodlands and marshes.

Zenobiella subrufescens	Local	Local

Known from nine or ten sites in the county with about five recent records. This is a species of generally damp, old woodland or marshes, which has a western distribution in Britain.

Terrestrial species which have been excluded

Arion flagellus)	Critical taxa which are difficult to identify
Arion hortensis seg.)	
Boettgerilla pallens -	Introduced recently (first British record 1972)
Columella edentula	Probably under recorded and moderately common
Oxychilus draparnaudi	Synanthropic species
Vertigo pygmaea	Probably commoner than records suggest
Pupilla muscorum	10 records, probably not uncommon in limestone dales

FRESHWATER SPECIES

Theodoxus fluviatilis	Local	Local

Six recent records from the Trent and Mersey Canal and near Cromford/Matlock Bath. Also recorded in the 19th century from the River Erewash. Known as the freshwater nerite, this species lives on stones and other hard surfaces in rivers and canals in England.

Viviparus viviparus	Local	Rare

Only record is from Whaley Bridge (probably the Peak Forest Canal and possibly in Cheshire) in 1969. The Peak Forest Canal was reopened to boats in the 1970s and mollusc surveys since then have indicated a major decline in mollusc populations; the current status of this species in the county is therefore uncertain and it may possibly have become extinct (pers. comm. I. F. Smith). This species is local but well-distributed in central Britain where it lives in canals, rivers and lakes.

	National Status	*Derbys. status*
Viviparus contectus	Local	Rare

Only recorded recently from Whaley Bridge (probably the Peak Forest Canal and possibly in Cheshire) but two late 19th century records from the Cromford Canal (SSSI) and the Trent and Mersey Canal. (See previous species for comments on the molluscs of the Peak Forest Canal). In Britain this is a local species of hard waters in canals and slow rivers; it is more tolerant of stagnant water than the preceding species.

Bithynia leachii — Local — Local

Recorded twice in the late 19th century from the east of the county, but only recently known from several stations along the Trent and Mersey Canal. A species of still or slowly flowing water in central and south England.

Aplexa hypnorum — Local — Not seen recently

Only known from three 19th century records in the county. A rather local but widely scattered species in England which is characteristic of ditches and other wetlands which seasonally dry up.

Planorbis planorbis — Local p

It is perhaps surprising that there are only two recent, and a further four old, records for this species because it is relatively common throughout most of south, east and central England. Consequently, it has been assigned provisional local status. It lives in a variety of waterbodies including quite small ponds, lakes, canals and slow-flowing rivers, generally in quite hard waters.

Gyraulus laevis — Nb — Not seen recently

There are two late 19th century records for this species, from the Mapperley and Willington areas. Shells of uncertain age have also been found in deposits at Creswell Crags. This is a very local but widely distributed species in Britain.

Pisidium supinum — Local — Local p

Known recently from two sites along the Trent and Mersey Canal, and from Whaley Bridge (possibly in Cheshire). It is a hard water species which is usually found in the mud of rivers and canals.

Pisidium hibernicum — Local — Rare p

First recorded in the county from the Cromford Canal and not since found elsewhere. Although rather local in Britain, this species is found in a wide range of waterbodies.

Pisidium moitessierianum — Nb — Local p

First recorded in the 1970s from three sites along the Trent and Mersey Canal and from the Erewash Canal. This species, our smallest bivalve mollusc, is a local species which is found in canals, lakes and rivers in central and southern England.

Note: the three *Pisidium* species are difficult to identify and are consequently infrequently recorded. Their distributions in the county are therefore uncertain so they have been assigned a provisional status.

Freshwater species which have been excluded

Sphaerium transversum	-	presence in county never confirmed
Lymnaea auricularia	-	probably not uncommon in south of county
Myxas glutinosa)		presence in county never confirmed
Lymnaea glabra)		

ACKNOWLEDGEMENTS

I am grateful to Dr. Michael Kerney, non-marine recorder of the Conchological Society of Great Britain and Ireland for making the Society's records available to me. It is primarily upon this information that the selection of the RDB species has been based. I am also grateful to Dr. L. Lloyd-Evans, Dr. R. C. Clinging, and Ian Smith who have provided further records and helpful advice.

REFERENCES AND FURTHER READING

BRATTON, J. H. ed (1991) *British Red Data Books 3. Invertebrates other than Insects.* Joint Nature Conservation Committee, Peterborough.

ELLIS, A. E. (1926) *British Snails.* Clarendon Press, Oxford. (Reprinted with appendix 1969)

ELLIS, A. E. (1962) Linnean Society, Synopses of the British fauna, No.13 *British Freshwater Bivalve Molluscs.* Linnean Society, London.

KERNEY, M. P. (1976) *Atlas of the non-marine Mollusca of the British Isles.* Institute of Terrestrial Ecology, Cambridge.

KERNEY, M. P. (1982) Vice-comital census of the non-marine Mollusca of the British Isles (8th edition) *Journal of Conchology*, 31: 63-71

KERNEY, M. P. & CAMERON, R. A. D. (1979) *A Field Guide to the Land Snails of Britain and North-West Europe.* Collins, London.

KERNEY, M. P. & STUBBS, A. (1980) *The Conservation of Snails, Slugs and Freshwater Mussels.* Nature Conservancy Council.

PFLEGER, V. & CHATFIELD, J.(1983) *A Guide to Snails of Britain and Europe.* Hamlyn, London.

TATTERSFIELD, P. (1990) Terrestrial Mollusc Faunas from some South Pennine Woodlands. *Journal of Conchology*, 33:355-374

TATTERSFIELD, P. (1991) Mollusc Faunas from some small Dark Peak Wetlands. *Sorby Record*, 28; 14-18

Further information about molluscs

For further information about the Conchological Society of Great Britain and Ireland (the national society dealing with all types of mollusc) contact: The Honorary Secretary, Mrs. S. Davies, 5 The Deans, Portishead, Bristol, BS20 8BG.

Theodoxus fluviatilis

INDEX
Scientific Names (Genera)

INTRODUCTION TO ARTHROPODS

FRED HARRISON

Listed under this name are a huge number of the world's smaller animal species including arachnids (spiders, mites, scorpions, etc.), crustaceans (crabs, shrimps, woodlice, etc.), insects, and the myriapods (centipedes and millipedes). The name refers to the jointed limbs and a body that is covered with a hard shell, or skeleton, and soft joints between the plates which enable the animal to move. The hard outer skeleton protects the body's vital organs and restricts water loss enabling species to spread into a wide range of habitats. Only the insects have the capability of true flight and this, coupled with their small size, has enabled them to fill ecological niches too small for the larger animals.

In Britain about 22,000 species of the insects alone have been named, of which (in round figures) 4,000 are beetles, 2,200 butterflies & moths, 5,200 two-winged flies and 1,650 true bugs. The population densities of these small animals is staggering, it has been estimated for instance that the total arthropod population in the top 30 centimetres of soil under grassland is 1400 million per acre of which 950 million were mites (Acarina) and 280 million springtails (Collembola) (see Russell 1957).

Mention wildlife conservation to most people and thoughts automatically turn to animals such as pandas, tigers, whales, etc. The larger and more visually attractive they are, then the more public sympathy is aroused and support gained in the fight for their survival. Generally speaking the number of naturalists in Britain who form the membership of societies that study arthropods, plants and birds, are roughly inversely proportional to the number of species in these groups. The large number of arthropod species, small in stature, unobtrusive by nature, with complex life histories, and inhabiting every conceivable niche (aquatic, subterranean, forest canopy, inside living and dead wood, etc.), ensures that entomologists must specialise and concentrate their studies upon a particular group, or small number of groups. The vast majority of entomologists are amateurs, and by far the greater number of these direct their interest towards butterflies and the larger moths which comprise less than 4% of the known arthropod fauna in Britain. As a result of this, the protection of habitats and their management are directed towards providing environmental conditions that appear to be suitable for ensuring the continuing survival of the larger animals, plants and perhaps butterflies. It is an admirable concept, and one that is, in the light of our current degree of knowledge, probably the only practical means of seeking to protect these species. Management works are, however, more often than not undertaken without taking into account the habitat needs of, and the effects upon, a vastly greater number of animal species whose presence and activities are essential to the survival of plants and the higher orders of animals. A predominant factor to be taken into account when seeking to ensure the protection of arthropod populations is that whilst some of them produce eggs which can remain dormant for several years (mainly crustaceans and aquatic species) until favourable conditions induce their hatching, most species have no mechanisms for surviving unfavourable conditions, and must complete one or more life cycles each year. The short life cycle, and inability to remain dormant during adverse conditions, plus the poor powers of dispersal exhibited by many species mean that it is crucial for their habitats to be suitable for each species at all stages of the life cycle, every year without fail.

Arthropods are small and largely regulate their body temperature through behavioural mechanisms, and their activity is largely influenced by climatic conditions. The physical structure of a habitat is a crucial factor in determining the potential range of micro-climatic conditions which they can exploit, and management activities such as grazing, mowing, tree felling, etc. modify this physical structure and thus the micro-climate. Choosing and applying the correct range of management options necessary to ensure the survival of the greatest variety of invertebrate fauna is not easy, particularly when management resources are often limited. No single type of grassland management for instance can maintain an entire arthropod fauna continuously in both space and time. A cessation of intensive management allows the phytophagous insects to increase, sometimes to a considerable degree for several years, but the abundance of some species then declines when management is not resumed. Mowing or burning brings about an immediate reduction in the height of vegetation, whilst grazing is more gradual, but all management techniques can have a catastrophic effect upon the arthropod fauna. Shorter grassland swards lead to a decrease in relative humidity and shade, and probably reduced micro-floral growth, which may explain the sensitivity of arthropod populations to the shortened sward length. Even the effects of cutting dead standing vegetation during the winter months has a severe effect because tall grassland provides important sites for over-wintering invertebrates. Site management operations which differ from those that already influence the vegetation of an area should not be introduced until a thorough survey of the invertebrate fauna has been undertaken and the consequences of proposed actions upon this fauna assessed.

Scallop shell moth

The creation of wildlife habitats following the reclamation of areas denuded of wildlife interest as a result of opencast coal extraction, industrial development, or even intensive farming, can bring about a significant increase in plant and animal populations, but should not be regarded as an alternative to the conservation of ancient habitats with their long established and rich assemblages of flora and fauna. Sites such as these are often important habitats, at least on a temporary basis, for those species that exploit ephemeral habitat conditions or resources, and derelict or recently reclaimed land frequently supports some of the largest populations of ruderal plants and invertebrates such as butterflies, but when site conditions are not maintained by management methods applied to arrest the progressive changes which take place in the soil structure and vegetation, then the decline of many species occurs and the local extinction of some species may take place.

It is possible to introduce plants and some invertebrates to such sites, once the appropriate conditions deemed suitable for their survival have been established, but it is not possible to re-introduce the myriads of arthropods which comprise the great majority of fauna in a long established habitat, because so little is known about their life histories and habitat requirements. Many arthropods feed upon dead organic materials, breaking these down often by acting in co-operation with fungi, and enabling nutrients to be recycled. Dead wood and accumulations of organic material (leaf litter, carrion and dung) are virtually absent from newly restored sites so a significant element of the arthropod populations environment is missing. The restored sites then, whilst appearing to be floristically rich, with large populations of some invertebrates, have in reality a greatly impoverished fauna and this is likely to have a considerable influence upon plant and animal communities living in these areas in the long term.

National Status

The British Red Data Books (see Shirt, 1987 & Bratton, 1991) define the national status of most threatened species and the criteria for the categories are as follows:

RDB1 — endangered

Species whose numbers have been reduced to a critical level or whose habitats have been so dramatically reduced that they are deemed to be in immediate danger of extinction.

Criteria
- species only known as a single population within one 10 km square
- species which only occur in habitats known to be especially vulnerable
- species which have shown a rapid and continuous decline over the last twenty years and now exist in five or fewer 10 km squares
- species believed extinct but if rediscovered would need protection

RDB2 — vulnerable

Species believed likely to move into the endangered category in the near future if the causal factors for their rarity continue operating. Included are species of which most or all of the populations are decreasing because of over-exploitation, extensive destruction of habitat or other environmental disturbance; species with populations that have been seriously depleted and whose ultimate security is not yet assured; and species with populations that are still abundant but are under threat from serious adverse factors throughout their range.

Criteria
- species declining throughout their range.
- species in vulnerable habitats
- species where populations are low

RDB3 — Rare

Species with small populations that are not at present endangered or vulnerable, but are at risk. These species are usually localised with restricted geographical areas or habitats, or are thinly scattered over a more extensive range.

Criteria - species which are estimated to exist in only 15 or fewer post-1960 10 km squares.

RDB4 — Out of danger

Species which were thought at some stage to be threatened but are now known to be more widespread and relatively secure due to conservation measures or removal of a previous threat.

RDB5 — Endemic

Species, sub-species, or distinct races which are not known to occur naturally outside of Britain.

Nationally Notable Species

In addition to the above there are lower categories of national rarity designated 'Nationally Notable', based upon those evolved by botanists, and three levels have been designated:

Notable A	(Na) – species having a very restricted national distribution. Known from 16-30 10km squares.
Notable B	(Nb) – species having a restricted national distribution. Known from 31-100 10km squares.
Notable N	(N) – uncertain at present whether Na or Nb should apply to species in this category.

It should be noted that although the nationally notable designations can be confidently applied to a few well recorded groups such as butterflies, the majority of arthropods are insufficiently known for species to be allocated to a particular category with the same degree of confidence, and most reviews of the scarcer species are regarded as provisional.

BIBLIOGRAPHY

1) Management and General

BRATTON, J. & ANDREWS, J. (1991) Invertebrate conservation principles and their application to broad-leaved woodland. *British Wildlife* No. 6, 335-344

Butterflies under threat team (1986) *The management of chalk grassland for butterflies* Focus on Nature Conservation No. 17, Nature Conservancy Council, Peterborough.

FRY, R. & LONSDALE, D. (1991) Habitat Conservation for Insects — a neglected green issue. *The Amateur Entomological Society*, Middlesex.

KIRBY, P. (1992) *Habitat Management for Invertebrates:* a practical handbook. JNCC & RSPB, Sandy, Bedfordshire.

MORRIS, M. G. (1978) *The effects of management on the fauna of grassland and scrub*, ITE Annual Report for 1977.

PURVIS, G. & CURRY J. P. (1981) The influence of sward management on foliage arthropod communities in a ley grassland. *Journal of Applied Ecology* 18, 711-725

RUSSELL, E. J. (1957) *The World of the Soil.* New Naturalist Series, Collins, London.

2) National Status

BRATTON, J. H. ed. (1991) *British Red Data Books 3: Invertebrates Other Than Insects.* Joint Nature Conservation Committee, Peterborough.

FALK, S. (1991) *A review of the scarce and threatened flies of Great Britain. Part 1.* Nature Conservancy Council, Peterborough. (Research and Survey in Nature Conservation no. 39).

HADLEY, Mark (1982) *A national review of British macro-lepidoptera.* Invertebrate site register report No. 100 Nature Conservancy Council, Peterborough.

HYMEN, P. S. rev. by PARSONS, M. S. (1992) *A review of the scarce and threatened Coleoptera of Great Britain* (Part 1) Joint Nature Conservation Committee (U.K. Nature Conservation no. 3).

HYMEN, P. S. rev. by PARSONS, M. A. (1994) *A review of the scarce and threatend Coleoptera of Great Britain* (Part 2). Joint Nature Conservation Committee (U.K. Nature Conservation no. 12).

KIRBY, P. (1992) *A review of the scarce and threatened hemiptera of Great Britain.* Joint Nature Conservation Committee, Peterborough. (U.K. Nature Conservation no. 21).

PARSONS, Mark (1984) *A provisional national review of the status of British micro-lepidoptera.* Nature Conservancy Council, Peterborough.

SHIRT, D. B. Ed. (1987) *British Red Data Books No. 2: Insects.* Nature Conservancy Council, Peterborough.

Ringlet butterfly

SPIDERS

STAN DOBSON

INTRODUCTION

Apart from the insects, arachnids constitute the largest group of arthropods. They are distinguished from other groups by having eight legs, no antennae or wings, simple eyes, and a life cycle comprising an egg stage then several instars to maturity. Spiders form the largest group within the arachnids and are distinguished from the others by having two segmented bodies and the ability to produce silk to use in a variety of ways.

Over 30,000 species of spider have been described world-wide, but the differences between them have not warranted a taxonomic division beyond suborder, that is, all spiders are in one order: Araneae. In this country, all except one species are in the suborder Labidognatha, and that one species, *Atypus affinis*, does not occur in Derbyshire.

With the possible exception of heavily polluted areas, spiders are found in all the available ecological niches present in Derbyshire, ranging from dense woodland to open moorland, from underground to treetops, and from total submersion in water to very dry domestic situations. They are to be found at all times of the year and can survive wide temperature ranges. The majority are comparatively short-lived, but some of the larger species may survive as adults for two or three years.

Spider Recording in Derbyshire

Unlike some of the adjoining counties, Derbyshire does not appear to have had any arachnologist actively working there before the late 1970s. J. R. Parker, one-time president of the British Arachnological Society, lived at Matlock for a few years, but at a period in his life when he had little time to pursue his interest in spiders. Apart from this, D. W. Mackie, who lived in Stockport for many years, made occasional excursions into the Peak District but published nothing. His records, and some of H. W. Freeston's made in the early 1900s, are to be found in Manchester Museum, and there are a few other casual records.

In 1977, the present author started collecting in Northwest Derbyshire and a considerable amount of data accumulated on the spiders in the High Peak and adjoining areas. This was augmented by pitfall trap material collected by P. Anderson from Harrop Moss, and very greatly by a number of field trips and a field course at Losehill Hall organised by M. J. Roberts during the 1980s, and a British Arachnological Society weekend at Losehill Hall during 1990.

All this resulted in a lot of information about spiders in the north of the county, but, with one exception, very little was obtained from other areas. The exception was a valuable input from an extensive programme of pitfall trapping carried out by M. Greenwood of Loughborough University at three sites near Repton. Recently, a useful input of records has come from mid and south Derbyshire due to a small number of people, principally F. Harrison, I. Viles, G. Maynard and R. Merritt, sending specimens of some of the more conspicuous spiders for identification. Although the numbers are comparatively small, these casual collections have resulted in several species being added to the county list. Further inputs have been pitfall trap material from two areas: the north east of the county obtained through Sheffield Museum and an invertebrate survey carried out by D. Cullen of the Manchester Metropolitan University at a reclaimed quarry site near Buxton, National Trust data which has been made available, and visits made by another arachnologist, J. D. Stanney, to previously unvisited sites. These have resulted in more county additions.

There are a number of species which appear on the distribution maps produced by Merrett (1974), with updates by Merrett (1975) and Merrett (1982), which have not been rediscovered. These records have no provenance; they were culled from Bristowe (1939) and other sources where the original data have been lost. They have, therefore, not been included in the main list, but are given in Appendix B.

Selection of Red Data Book species

At the time of writing (August 1995), the number of species recorded in Derbyshire stands at 293 (there are three other doubtful ones which have been square bracketed in the list), but because of the sparse coverage of the south of the county as outlined above, this is certainly not the true figure. Applying the Red Data Book criteria to this list results in 182 candidates. In fact, only the criterion of number of sites need be applied since the other criteria are subsumed within this. For ease of application, site has been largely interpreted as 1km square, since there are very few cases where two sites occur in the same square or, conversely, where a square boundary divides a site.

In order to reduce the problem and to make the selection of species for inclusion more realistic, the site criterion has been modified to not more than 4. The reason for this is that the species which are now excluded may all be judged to be more or less common and/or widespread nationally and in adjoining counties, whereas some (but by no means all) of those found at four sites or less may be considered to be uncommon or restricted in distribution. Applying this new criterion reduces the candidates to 121. Those excluded are listed in Appendix A, the number of sites being given after each name. It is inevitable that all these will achieve the requisite number of sites within the next few years, as will many of those in the main list when more work is carried out in other parts of the county. It is ironic that the majority of the species which should be listed here have almost certainly not yet been discovered!

After application of the site criterion, the species list which results has been split into two. The first (23 species) consists of spiders which are under-recorded but which are not believed to be under threat; the second (98 species) consists of spiders which may be under-recorded but which may only occur in small numbers. The reasons for being under-recorded are that they are (1) domestic species found exclusively inside houses leading to difficulties in obtaining records, (2) species which probably avoid high ground and which are more likely to be encountered outside the Peak District where most recording has been done, and the largest group, (3) small and/or inconspicuous species only found by deliberate searching in suitable places.

Roberts (1985 & 1987) has been used to provide the indications in the text of the national distribution of some of the species. The National Status is taken from the list supplied by the Joint Nature Conservation Committee dated 18 November 1992. It should be borne in mind that the terms Local and Common may not apply at county level.

Nomenclature follows Merrett, Locket and Millidge (1985) with amendments given in Merrett and Millidge (1992).

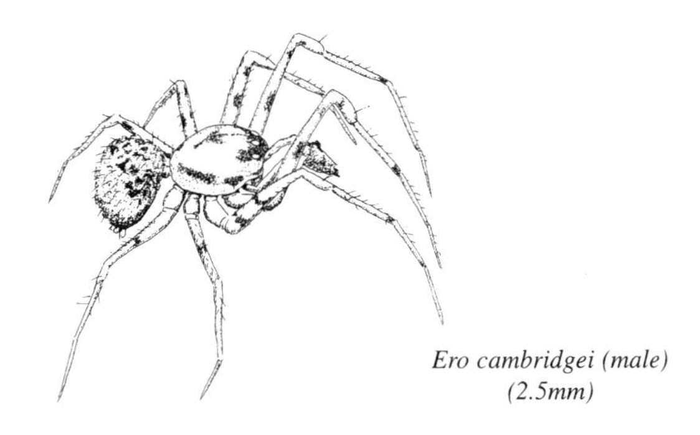

Ero cambridgei (male)
(2.5mm)

Spiders which are under-recorded but which are not believed to be under threat

	National status	*Post-1979 sites*
Oonops domesticus	Synanthropic	2

This tiny pink spider is found virtually exclusively indoors. It is very inconspicuous, building no web and appearing only after dark. It is probably quite common but so easily overlooked that it is no surprise to find that the only recent records are from Birch Vale and Glossop in the houses of the only two members of the British Arachnological Society who now live in the county, and there is one old record from Matlock where a member used to live!

Scotophaeus blackwalli	Synanthropic	2

A domestic species which is possibly not uncommon in the county but under-recorded. Although fairly large and conspicuous, it does not make a web and only appears after dark, usually indoors. Most people would not recognize it as different from other spiders which wander about inside houses at night. One record from Alfreton Hall (found during a DWT meeting!) and one from New Mills.

Zelotes latreillei	Local	3

Probably the commonest member of the genus, this spider, which lives under stones, has been found in Wye Dale, Ladybower Wood and Lathkill Dale; it should be found elsewhere in future.

Phrurolithus festivus	Common	2

Found once at Pleasley and in a pitfall trap at Repton. This is a sunshine loving species which may explain its absence from the High Peak. Status unknown.

	National status	*Post-1979 sites*
Zora spinimana	Common	2

Recorded from Goyt Valley, Crich Chase and Ladybower Wood. Generally not uncommon and should be more widespread.

Philodromus aureolus	Common	3

This very common species has, amazingly, only been recorded from Hilton Gravel Pits and Cromford Canal Reserve.

Ero cambridgei	Common	3

Recorded from Chee Dale, Ravensdale and Long Eaton.

Ero furcata	Common	2

Recorded from Goyt Valley and Monks Dale with an old record from Chee Dale.

Theridion varians	Common	2

A fairly common spider recorded only from Birch Vale and Erin, near Poolsbrook.

Theridion pallens	Common	4

A small spider which can be abundant in woodland, particularly on oaks. Recorded from Monks Dale, Ravensdale, Ladybower Wood and Highlow Wood.

Theonoe minutissima	Local	4

This is possibly the smallest nonlinyphiid spider and, as such, it is easily overlooked. It has been recorded from Birch Vale, Chee Dale, Ladybower Wood and Ringinglow Bog, and will doubtless turn up more frequently in future.

Araniella cucurbitina	Common	3

This small green orbweaver has so far only been recorded from Birch Vale, Hilton Gravel Pits and Poolsbrook.

Walckenaeria cucullata	Common	3

Recorded from Long Clough, Derwent Valley and Ladybower Wood.

Entelecara erythropus	Local	3

Recorded recently from Birch Vale, Glossop and Ladybower Wood. Old records from Derby, Derwent reservoir and Matlock.

Moebilia penicillata	Common	2

Recorded from Ladybower Wood and Hilton Gravel Pits.

Gnathonarium dentatum	Common	3

Recorded from Watford Lodge, Monks Dale and Long Eaton.

	National status	Post-1979 sites
Peponocranium ludicrum	Common	3

Recorded from Kinder Bank Wood, Ladybower Wood and Great Rocks Dale.

Pocadicnemis juncea	Common	3

Fairly recently separated from the common *P. pumila* and possibly as common when more records are received. Recorded from Repton, Spinkhill and Great Rocks Dale.

Tapinocyba praecox	Local	2

Recorded from Long Clough and Ladybower Wood. One old record from Stanton Moor.

Aphileta misera	Local	4

Recorded from Goyt Valley, Birch Vale, Kinder and Beeley.

Centromerus arcanus	Local	3

A spider of high ground, recorded from Goyt Valley, Monks Dale and Griffe Grange Farm Wood.

Tallusia experta	Common	4

Recorded from Long Clough, Alport Dale, Ringinglow Bog and Norbriggs Marsh, Staveley.

Kaestneria pullata	Common	4

Recorded from Long Clough, Ravensdale, Hilton Gravel Pits and South Wingfield.

Spiders which may be under-recorded but which may only occur in small numbers.

	National status	Post-1979 sites
Amaurobius ferox	Common	2

Shown on the Merrett distribution maps, but only two recent records, from Sandiacre and Winster. Common in some parts of the country, it is difficult to predict the status in Derbyshire.

Dictyna uncinata	Common	1

Locally common in some parts of the country, this species has only been recorded from Spinkhill.

	National status	*Post-1979 sites*
Pholcus phalangioides	Synanthropic	3

Very common indoors in the southern counties of England and Wales, this spider thins out northwards, but appears to be increasing its range, presumably due to the movement of people with their furniture. There is a small colony at Losehill Hall and individuals have been found at Birch Vale and Low Leighton; doubtless others will turn up in future.

Psilochorus simoni	Synanthropic	0

An inhabitant of dry cellars. There is only one old record from a wine cellar in a hotel at Matlock.

[Drassodes lapidosus]		0

Single records from Sheldon and Kinder. These records cannot be checked and Roberts (1985) says that many specimens recorded as *D. lapidosus* have turned out to be *D. cupreus* which is more common. Misrecording may arise because, at one time, *D. cupreus* was considered to be a subspecies of *D. lapidosus.*

Zelotes subterraneus	Local	3

This has only recently been separated from *Z. apricorum* so the status and distribution is unclear. Nationally, it appears to be much less common than the latter species; but in Derbyshire, it would appear to be found almost as frequently. It has been recorded from Chee Dale, Ravensdale and Lathkill Dale, in the first two sites alongside *Z. apricorum.* This might turn out to be a Derbyshire speciality requiring site conservation.

Gnaphosa leporina	Local	1

Only one of this uncommon species has been found, in a pitfall at Great Rocks Dale.

Clubiona corticalis	Common	1

A southern species which thins out northwards. There is no reason why it should only have been recorded once, but it has only been found in a pitfall trap at Repton.

Clubiona stagnatilis	Common	1

A nationally common species of wet habitats with, surprisingly, only one record from a pitfall trap at Erin, near Poolsbrook.

Clubiona neglecta	Local	2

Records from Rose End Meadows and Great Rocks Dale.

Scotina celans	Local	1

An inconspicuous species which has only been found once at Ravensdale. Its status in the county is uncertain; it may be rare or it may be under-recorded.

	National status	Post-1979 sites

Xysticus erraticus Local 1

Old records from Deep Dale and Dovedale and one recent pitfall record from Great Rocks Dale.

Ozyptila atomaria Common 2

Recorded only from Lathkill Dale and Great Rocks Dale. The species may be decreasing.

Philodromus dispar Common 3

Near the north of its range, this arboreal species has been recorded from Rose End Meadows, Walton and indoors at Birch Vale.

Salticus cingulatus Common 2

Two records only of this woodland species from Hilton Gravel Pits and Spinkhill.

Heliophanus cupreus Common 0

Although this species can coexist with *H. flavipes*, it seems to be largely replaced by the latter in the dales. One old record only, from Ravensdale.

Euophrys erratica Local 2

Recorded from Chee Dale and Ravensdale.

Euophrys aequipes Local 1

This tiny jumping spider is found mainly in the south. One from a pitfall in Great Rocks Dale.

Euophrys lanigera Local 1

An uncommon southern species which appears to be increasing in numbers and extending its range northwards. One recent record from Sandiacre.

Sitticus pubescens Local 1

A semi-domestic species which tends to occur in and around houses. It may be much more common than is indicated by the single site at Birch Vale, where it is found regularly, and one old record from Matlock.

Pardosa agricola Local 1

One record from Lathkill Dale. Status unknown.

	National status	*Post-1979 sites*
[Pardosa monticola]		

One old record from Kinder Scout, and another more recent from Sheldon. There is a possibility of confusion with *P. palustris*, but checking may be impossible.

Alopecosa barbipes	Common	1

One record from Great Rocks Dale.

Pirata hygrophilus	Common	2

A common wolf spider of wet bogs, which has only been found at one colony in Long Clough and a single record from Cromford Canal. It could occur elsewhere in similar habitats.

Pirata latitans	Local	1

Recorded only from Rose End Meadows. Probably has a very limited distribution in the county.

Argyroneta aquatica	Local	2

The water spider could occur in any pond or slow stream in the county but, so far, it has only been recorded from the canal at Cromford and a pond at Long Eaton. The problem with this species is that it has to be searched for as a dipping exercise, rather than a general spider hunt. Conservation will consist solely of habitat protection.

[Tegenaria atrica]

Since *T. saeva* and *T. gigantea* were collectively known as *T. atrica* before they were separated, the identity of an old record from Killamarsh is uncertain.

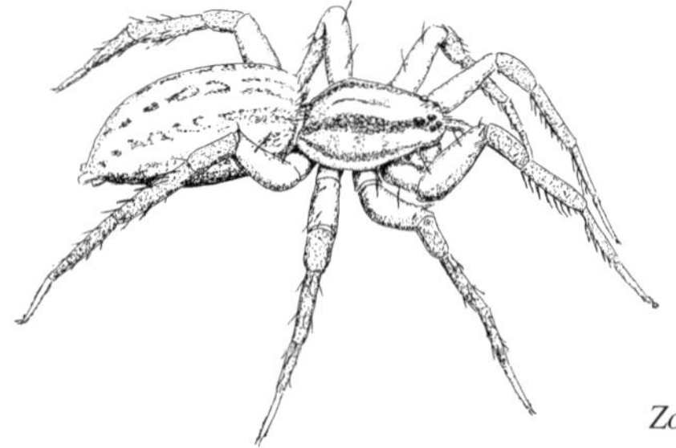

Zora spinimana (female)
(6mm)

	National status	Post-1979 sites

Tegenaria agrestis Local 1

An outdoor species, unlike many other of the Tegenaria group, which tends to be found on waste sites. One old record from Derby might confirm this, but it has recently been found in abundance at a reclaimed quarry site at Great Rocks Dale.

Tegenaria silvestris Local 1

Another outdoor species found once at Great Rocks Dale.

Hahnia nava Local 0

Surprisingly, only one old record from Breadsall Moor.

Hahnia helveola Local 2

Recorded only from Ravensdale and Lathkill Dale.

Euryopis flavomaculata Local 1

Kinder Bank Wood seems to be the only known site in the county for this attractive and quite rare little spider.

Anelisomus vittatus Common 2

Found on trees, this widespread species has, surprisingly, only been recorded twice, from Chesterfield and Chesterfield Canal.

Achaearanea tepidariorum Synanthropic 0

A large theridiid which normally survives in heated greenhouses. It has been recorded once from Matlock.

Achaearanea simulans Nb 1

A surprise find in a pitfall trap at Erin, near Poolsbrook, turned out to be the most northerly record of this uncommon spider.

Theridion impressum Local 1

One record from Stoney Clouds Nature Reserve, near Erewash.

Theridion melanurum Synanthropic 1

Normally associated with buildings, recorded only once from Long Clough, habitat unknown.

Theridion mystaceum Common 3

Recorded from Birch Vale, Alport Dale and Ladybower Wood.

	National status	*Post-1979 sites*
Theridion tinctum	Local	1

A southern species thinning out northwards. One record from Shipley Country Park.

Rugathodes bellicosus	Local	1

This high-ground species has only been taken in pitfalls at Great Rocks Dale.

Enoplognatha thoracica	Local	1

Found under stones in suitable places, only recorded once from Monks Dale.

Robertus neglectus	Local	2

An uncommon species, recorded from Monks Dale and Fenny Bentley and one old record from Grin Wood, Buxton.

Tetragnatha obtusa	Common	3

This spider has recently been rediscovered at Cromford Canal, Hilcote and Oakerthorpe.

Zygiella atrica	Common	1

This spider has recently been rediscovered at Trent Meadows.

Tetragnatha striata	Nb	2

This rare spider, found almost exclusively in reed-beds at twenty or so sites in Britain, was discovered by R. Merritt at two widely separated sites: Long Eaton and Alfreton.

Hypsosinga pygmaea	Local	2

An uncommon spider recorded only from Spinkhill and Erin, near Poolsbrook.

Cercidia prominens	Local	1

Another uncommon spider recorded only from Crich Chase.

Cyclosa conica	Local	1

A species of dark woodland. Recorded from Clough Wood.

Walckenaeria nodosa	Local	2

Recorded from Birch Vale and Great Rocks Dale.

Walckenaeria atrotibiale	Local	1

Recorded from Kinder.

Walckenaeria incisa	Nb	1

Recorded from Kinder.

	National status	Post-1979 sites

Walckenaeria dysderoides — Local — 2

Recorded from Ogston Woodlands and one pitfall trap record from Ladybower Wood.

Walckenaeria furcillata — Local — 2

Recorded from Monks Dale and Ravensdale.

Walckenaeria unicornis — Common — 1

Recorded from Erin, near Poolsbrook.

Dicymbium brevisetosum — Local — 3

Recorded from Beeley, Broadhurst Edge Wood and Birch Vale.

Entelecara errata — Nb — 1

Recorded from Ladybower Wood.

Tmeticus affinis — Local — 1

Recorded from Repton and Kinder.

Hypomma cornutum — Common — 2

Recorded from Clough Wood and Kedleston Park.

Baryphyma pratense — Local — 2

Recorded from Repton and Sawley Oxbow.

Hypselistes jacksoni — Local — 1

A rare spider recorded once from Ringinglow Bog.

Pelecopsis parallela — Local — 1

Recorded from Kinder.

Silemetopus reussi — Local — 1

Recorded from Great Rocks Dale.

Mecopisthes peusi — Nb — 1

Recorded from Ladybower Wood.

Evansia merens — Local — 1

Normally found with ants. Recorded from Kinder Bank Wood. Common in the north but thinning out southwards.

	National status	*Post-1979 sites*
Troxochrus scabriculus	Local	2

Recorded from Repton (two sites).

Minyriolus pusillus	Common	2

Recorded from Goyt Valley (two sites).

Thyreosthenius parasiticus	Local	2

Recorded from Ogston Woodlands and Wensley with one old record from Chisworth.

Thyreosthenius biovatus	Local	1

Normally found only in ants nests. Recorded from Alport Dale.

Monocephalus castaneipes	Local	2

Recorded from Lathkill Dale and Ladybower Wood.

Micrargus apertus	Local	2

Only fairly recently separated from the common *M. herbigradus*, this species will probably turn up more frequently although evidence suggests that it is much less common. Recorded from Chee Dale and Ladybower Wood.

Micrargus subaequalis	Local	2

Recorded from Repton and Rose End Meadows.

Araeoncus humilis	Common	1

Recorded from Kinder Bank Wood.

Araeoncus crassiceps	Local	1

Recorded from Birch Vale.

Typhochrestus digitatus	Local	1

Recorded from Great Rocks Dale with one old record from Kinder.

Milleriana inerrans	Local	3

Recorded from Repton (two sites) and Great Rocks Dale.

Erigone longipalpis	Local	2

A spider of wet habitats, more common on the coast. Recorded from Repton and, surprisingly, from Great Rocks Dale.

	National status	Post-1979 sites
Erigone welchi	Na	1

This is probably the rarest spider so far discovered in Derbyshire. The typical habitat is *Sphagnum* bog on high ground where it spins its web just above the water surface, and it has so far been discovered in about ten localities scattered throughout Britain. It has been recorded once from Mermaids Pool, Kinder, where it is probably well established.

Prinerigone vagans Local 1

Recorded from Monks Dale.

Leptothrix hardyi Local 2

Recorded from Harrop Moss, Kinder and Great Rocks Dale.

Halorates distinctus Local 3

Recorded from Repton (three sites).

Asthenargus paganus Local 1

A rare species, recorded from Alport Dale.

Jacksonella falconeri Local 1

Recorded from Kinder (three pitfall trap sites).

Ostearius melanopygius Naturalised 0

One old record from Derby. This is surprising as this species is supposed to be increasing its range.

Porrhomma egeria Local 1

A cave-dweller, recorded from Cumberland Cavern and Robin Hoods Cave, Cresswell Crags.

Porrhomma montanum Local 2

Found normally on high ground, recorded from Goyt Valley and Kinder (two pitfall trap sites).

Agyneta subtilis Common 1

Recently re-discovered at Great Rocks Dale.

Agyneta ramosa Local 1

Recorded from Hilton Gravel Pits.

Meioneta innotabilis Common 1

Recorded from Ladybower Wood.

	National status	*Post-1979 sites*
Meioneta mossica	Unknown	1

This spider has only recently been separated from *M. saxatilis* and its status is unknown. It has been taken in pitfalls on Kinder.

Meioneta beata	Local	2

Recorded from Monks Dale and Ravensdale.

Meioneta gulosa	Local	2

Recorded from Kinder and Great Rocks Dale.

Maro minutus	Local	1

Recorded from Goyt Valley.

Saaristoa firma	Local	1

Recorded from Monks Dale.

Kaestneria dorsalis	Common	0

Surprisingly, only one pre-1980 record from Birch Vale.

Floronia bucculenta	Local	1

Recorded from Great Rocks Dale.

Lepthyphantes nebulosus	Common	1

A species found in houses, more common in the south. Recorded only from New Mills.

Lepthyphantes tenebricola	Local	1

A not uncommon species only recorded from Ladybower Wood. Old records from Lathkill Dale and Tideswell Dale.

Lepthyphantes expunctus	Local	2

This species is common in pine woods in parts of Scotland. Records from two sites in the Goyt Valley are the southernmost records for Britain.

Pityohyphantes phrygianus	Na	1

This spider appeared some years ago in the northeast and has slowly expanded its range in Scotland and northern England. It is found in conifer plantations. One record from a pitfall trap in Alport Dale indicates that it may be invading Derbyshire, but there have been no further records.

	National status	*Post-1979 sites*
Microlinyphia impigra	Local	2

An uncommon spider of wet places. Recorded from Poolsbrook and Long Eaton.

Allomengea vidua	Local	3

Recorded in some numbers from pitfall trapping at three sites at Repton.

APPENDIX A

Species recorded from more than four, but less than ten, post 1979 sites.

Species	Sites	Species	Sites
Dictyna arundinacea	9	*Meta menardi*	7
Zelotes apricorum	6	*Ceratinella brevipes*	8
Micaria pulicaria	6	*Dicymbium nigrum*	8
Clubiona phragmitis	5	*Dicymbium tibiale*	9
Clubiona comta	8	*Metopobactrus prominulus*	5
Clubiona brevipes	5	*Gonatium rubellum*	5
Clubiona trivialis	5	*Oedothorax agrestis*	9
Agroeca proxima	5	*Oedothorax apicatus*	5
Ozyptila trux	7	*Tapinocyba pallens*	9
Philodromus cespitum	9	*Gongylidiellum vivum*	7
Heliophanus flavipes	9	*Scotinotylus evansi*	5
Neon reticulatus	9	*Diplocentria bidentata*	5
Pardosa palustris	8	*Porrhomma convexum*	5
Pardosa nigriceps	9	*Porrhomma pallidum*	6
Pardosa lugubris	8	*Centromerus sylvaticus*	7
Trochosa ruricola	6	*Centromerus prudens*	6
Pisaura mirabilis	7	*Bathyphantes approximatus*	6
Tegenaria gigantea	9	*Bathyphantes parvulus*	8
Tegenaria saeva	5	*Lepthyphantes leprosus*	8
Tegenaria domestica	5	*Lepthyphantes obscurus*	9
Antistea elegans	8	*Lepthyphantes cristatus*	9
Steatoda bipunctata	8	*Lepthyphantes flavipes*	6
Theridion bimaculatum	9	*Lepthyphantes pallidus*	9
Nesticus cellulanus	9	*Allomengea scopigera*	8

APPENDIX B

Species shown in British Spiders, vol 3, as being recorded from Derbyshire, but no recent records.

Misumena vatia
Xysticus audax
Marpissa muscosa
Evarcha falcata
Trochosa spinipalpis
Pirata piscatorius
Agelena labyrinthica
Hahnia pusilla
Theridion pictum
Araneus marmoreus
Gongylidiellum latebricola
Latithorax faustus
Agyneta cauta

REFERENCES

BRISTOWE W. S. (1939). *The Comity of spiders*. Vol 1. Ray Society.

MERRETT P. (1974). Distribution maps of British spiders. *British Spiders*. Vol 3. Locket, Millidge & Merrett. Ray Society.

MERRETT P. (1975). New county records of British spiders. *Bulletin, British Arachnological Society*, 3(5).

MERRETT P. (1982). New county records of British spiders. *Bulletin, British Arachnological Society*, 5(7).

MERRETT P., LOCKET G. H. & MILLIDGE A. F. (1985). A check list of British spiders. *Bulletin, British Arachnological Society*, 6(9).

MERRETT P. & MILLIDGE A. F. (1992). Amendments to the check list of British spiders. *Bulletin, British Arachnological Society*, 9(1).

ROBERTS M. J. (1985 & 1987). *The spiders of Great Britain and Ireland*. Vols 1 & 2. Harley Books, Colchester.

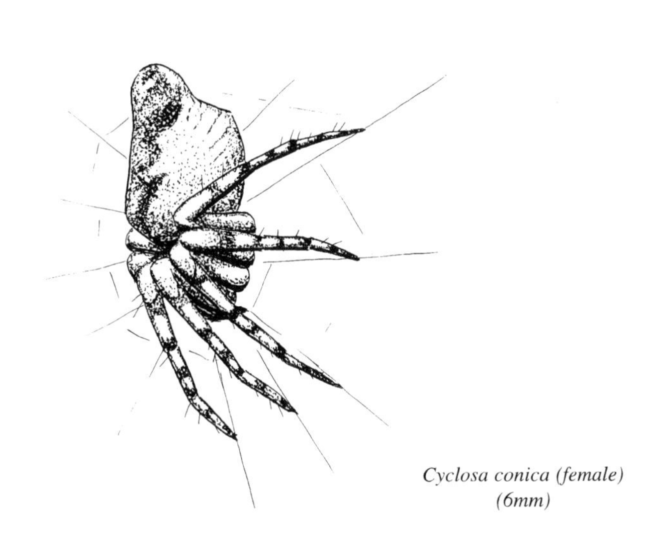

Cyclosa conica (female)
(6mm)

INDEX
Scientific Names (Genera)

CRUSTACEANS

PAT BRASSLEY

No. of post-1980
1 km squares

Crayfish *Austropotamobius pallipes* 10+

The native crayfish is a protected species under the Wildlife and Countryside Act (1981) and, internationally, under the Berne Convention.

The crayfish, found in hard water derived from limestone, was relatively common in Derbyshire. It is threatened by low river flows, pollution and introduced non-native species which have brought in crayfish plague. There are reports of the plague affecting crayfish in the River Wye (pers. comm. 1994).

Crayfish

DRAGONFLIES

RODERICK DUNN

Dragonflies belong to the order Odonata. World-wide there are about 5000 extant species with 114 found in Europe and 39 species surviving in Great Britain and Ireland. Sadly, after a long and successful history, the extant species of dragonfly face increasing threats to their survival with 3 species having disappeared from England since 1951. To date, 20 species have bred in Derbyshire giving us about 50% of the national list.

There are two suborders of Odonata, the Zygoptera (damselflies) and the Anisoptera (dragonflies proper). The term dragonfly is used generally to encompass both orders. There is in fact a third suborder (Anisozygoptera) but this relict suborder dating back at least to the Triassic, has species surviving only in Japan and the Himalayas.

Derbyshire can lay claim to the earliest dragonfly fossils yet found with *Erasipteron bolsoveri*, being discovered in 1978 a thousand metres beneath the surface of Bolsover Colliery, (Whalley, 1978) and *Tupus diluculum* also from the same mine. These were among the very first creatures to develop flight, preceding pteranodons by 100 million years and birds by 150 million years. The dragonflies that fly today, some 300 million years later, are very similar in structure to these primeval swamp dwellers of the Carboniferous period and it is not surprising that in some circles dragonflies are referred to as 'flying fossils'.

Dragonflies are large, beautiful insects, quite distinctive from other insects. An efficient body design together with exceptional flying skills and a well developed sense of sight are the key factors to the success of the dragonfly surviving major climatic and hence environmental changes. Dragonflies undergo incomplete metamorphosis (no pupal stage) and the easily observed transformation from ungainly larva to winged adult is one of the wondrous sights of nature. Although dragonflies have always been aerial insects throughout their evolution, today they are true aquatic insects spending 98% of their life cycle as underwater larvae. The adults live barely a month (predators willing) during which they breed — the females laying eggs directly into the water or inserting them into aquatic vegetation.

Dragonflies may have survived three hundred million years but their future is uncertain. By requiring aquatic habitats in which to mature, dragonfly larvae are vulnerable and their presence or otherwise at water bodies is often an indication as to the health of the water. Dragonflies breed in natural habitats ranging from rivers, streams, ponds, ox-bow lakes, marshes and bogs and in man made habitats such as reservoirs, canals, dew ponds, disused gravel pits, ditches, garden ponds and ornamental lakes. Some species are able to breed successfully in a variety of habitats, but others are specific to a particular habitat and are hence more at risk. The distribution of dragonflies in Derbyshire can be related to the topography which, generalised, can be split into four contrasting regions: the upland north supporting blanket bog on the Millstone Grit; the limestone plateau in the west; the lowland Coal Measures in the east and, in the south, a lowland region of Keuper Marls, sandstones and pebblebeds. The pools on the blanket bogs support few species; the absence of standing water on limestone leaves this area almost devoid of species and the rivers on the Coal Measures support little species diversity.

The greatest number of species (16 of the 20) are found in the south of the county which, apart from having the greatest variety of habitats, is also the geographic and climatic limit for the majority of the national species. There are occasional exceptions to these generalisations. The threats to wetland habitats nationally have been well documented and the dragonfly sites in Derbyshire are equally at risk due to pollution, infill, 'improvements', drainage, property development and, sadly, incorrect or deficient site management. 90% of farm ponds have disappeared in the last 30 years and some rivers remain heavily polluted. Ironically, we have the most polluted river in the country (the Rother) and also the cleanest (the Lathkill) and neither support dragonflies: the Rother for obvious reasons and the Lathkill due to the overstocking of fish which are heavily predacious on dragonfly larvae. The conversion of natural dragonfly hinterland feeding grounds to grass monoculture or other development also influences dragonfly colonies and populations. Although we have half the national list breeding in the county, some species exist at only a few isolated sites and their future presence in the county cannot be guaranteed.

Dragonfly Recording in Derbyshire

Dragonflies can be observed on the wing in Derbyshire from May to October but they can be recorded as larvae throughout the year. Nationally, Derbyshire is not an exceptional county for dragonflies and this possibly reflects the uneven history of recording in the county. Although observations of dragonflies have been taking place since the nineteenth century, few corroborated records survive today. Lucas (1900) does not list a single species for the county even though the Victoria County History (VCH) for Derbyshire (Eaton, 1905) mentions that Brown (1863) lists ten species. Unfortunately, Brown's list is also duplicated in the VCH of Staffordshire and therefore it cannot be guaranteed which species were specific to Derbyshire. Longfield (1949) lists 15 species. Now the criteria for a corroborated record include grid reference, date and recorder and, for a breeding record, confirmation of successful breeding either by the presence of larvae or exuviae (larval skins). Oviposition (egg laying) is not regarded as evidence of successful breeding. Dragonflies are exceptional fliers and are not always observed at their specific breeding sites. In the early 1980s there was an influx to the county of a number of recorders with a special interest in the group and by 1987 six new species had been confirmed to have bred in the county. Nationally, there is some evidence to suggest that a number of species are extending their range northwards and an increase in observations of some species in the south of the county may reflect this.

Platycnemis pennipes and *Calopteryx virgo* are not included in the county list as the sightings have been proven to be of mistaken identity and/or the records are uncorroborated.

Of the six species included here, all are Anisopterans and none qualifies for the national RDB either as Category 1, 2, or 3. However, one species is classified as Na — the nationally scarce category. All, with the exception of this species, are common nationally. The criteria considered for inclusion here are combinations of the following: confirmed breeding at less than ten sites since 1977; the vulnerability of these sites to pollution, development or lack of management; the distribution of individual colonies of the species (recolonisation potential); whether the species is on the limit of its natural range coupled with one or more of the above factors.

Each of the six species has been placed in one of three categories of paucity in the county:

Rare	-	found from only one breeding site
Scarce	-	found from less than five breeding sites
Local	-	found from less than ten scattered breeding sites

Using these guidelines, we have two rare, three scarce and one local species of dragonfly in Derbyshire. Nomenclature follows Hammond (1983) and arrangement Davies (1984,1985).

National Status *Derbys. Status*

The Emperor Dragonfly *Anax imperator* 2

A species which seems to be extending its range nationally. Larvae and exuviae can be found annually in the south of the county. It is a casual visitor elsewhere. The species prefers ponds and lakes with abundant floating aquatic vegetation on which the females land to oviposit. Breeds at one Derbyshire Wildlife Trust reserve and visited another in 1994.

The Golden-ringed Dragonfly *Cordulegaster boltonii* 1

One main breeding site is known, a stretch of moorland stream near Baslow. Adults also frequent other water bodies. Males patrol a section of territory at acid streams that have sluggish, silted areas where females oviposit and the larvae develop concealed in the silt.

The Four-spotted Chaser *Libellula quadrimaculata* 3

Main breeding strongholds near Grindleford. A great wanderer and the individuals sighted in the south of the county could have come from anywhere. Prefers to breed in acid pools that support some emergent vegetation.

The Black-tailed Skimmer *Orthetrum cancellatum* 3

Has bred at three sites but observation on an annual basis cannot be guaranteed. Breeds in ponds, lakes and gravel pits that have some marginal vegetation and, ideally, sandy banks or adjacent paths on which to bask. A species which has extended its range into south Derbyshire.

The Ruddy Darter *Sympetrum sanguineum* 8

Has bred at eight sites. Main stronghold is at a Derbyshire Wildlife Trust reserve in the south of the county. This dainty species prefers still waters that are well reeded or almost choked with appropriate aquatic vegetation. Some new sites near Chesterfield in 1992.

The White-faced Darter *Leucorrhinia dubia* Na 1

One site near Matlock. An acid water species which requires an abundance of surface and, importantly, underwater *Sphagnum* and macrophytes for the larvae to mature successfully. A nationally scarce species under great threat in Derbyshire due to pollution of habitat and subsequent demise of the submerged *Sphagnum*. Numbers have steadily declined since 1987 and Merritt (pers.comm.) postulates that the species may have been introduced into the county in the 1980s. No exuviae or adults seen since 1991.

ACKNOWLEDGEMENTS

I am indebted to Robert Merritt for providing the VCH; Fred Harrison for some historical data; Nick Moyes at the County Museum for evaluating a suspect species and all who have contributed to the county recording scheme.

REFERENCES

BROWN, E. (1863). Fauna & Flora of the District surrounding Tutbury and Burton-on-Trent. In MOSLEY, O. *The Natural History of Tutbury*, London. pp.172-173

DAVIES, D. A. L. (1984). *The Dragonflies of the World:* a systematic list of the extant species of Odonata. Vol. 1 Zygoptera, Anisozygoptera. Societas Internationalis Odonologica Rapid Communications (Supplements) No. 3. 127pp.

DAVIES, D. A. L. (1985). *The Dragonflies of the World*: a systematic list of the extant species of Odonata. Vol. 2. Anisoptera. Societas Internationalis Odonologica Rapid Communications (Supplements) No. 5. 151pp.

EATON, A. E. (1905). *Odonata* in Page, W. (ed.) *Victoria History of the County of Derby* Vol. 1. Constable, London. pp.55-56.

HAMMOND, C. O. (1983). *The Dragonflies of Great Britain and Ireland.* 2nd Edition (revised by R. Merritt). Harley Books, Colchester. 116pp.

LONGFIELD, C. (1949). *The Dragonflies of the British Isles.* 2nd Edition, Warne & Co. London and New York. 256pp.

LUCAS, W. J. (1900). *British Dragonflies (Odonata).* Upcott Gill, London. 356pp.

RICHARDS, A. W. (1947). Additions to the Derbyshire List of Odonata. *The Entomologist.* Vol. 80, p.272.

RICHARDS, A. W. (1950). *Anax imperator* (Leach) in Derbyshire. *The Entomologist.* Vol. 83, p.238.

WHALLEY, P. (1978). Derbyshire's darning needle. *The New Scientist.* 15 June. 740-1

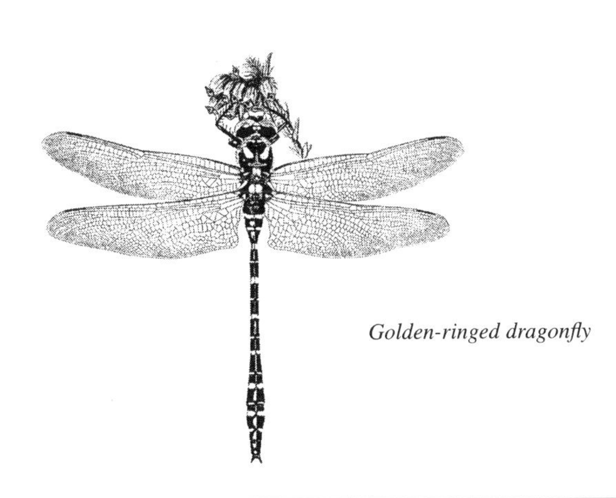

Golden-ringed dragonfly

GRASSHOPPERS AND CRICKETS

ROY FROST

Grasshoppers and crickets (Orthoptera) are generally large conspicuous insects, with long hind limbs giving them the ability to jump impressive distances in relation to their size, and having stridulatory mechanisms enabling each species to make characteristic sounds. Grasshoppers are predominantly diurnal and vegetarian; they have short robust antennae, and stridulate by rubbing their hind legs against their wings. Crickets and bush-crickets are more omnivorous and often nocturnal in habit, have long thread-like antennae, and produce sounds by rubbing the bases of their wings together.

Of the 27 species native to Britain, only eight have so far been confirmed in Derbyshire. This is primarily because orthopterans are sun and warmth-loving insects, so generally the further north one travels in Britain, the fewer species are present and the more restricted in range they become. However, the group has not until recent years been intensively studied locally and there has been little published material until the 1970s. It is possible that two or three additional species may occur undetected at a low density in the warmer parts of the county.

Generally they may be found in areas of ungrazed or lightly grazed grasslands, wasteland including scree, or open woodland and woodland-edge habitats. Since most, if not all, of these habitats have declined in the county, the overall populations of resident grasshoppers and crickets have suffered a similar decline.

Of the seven species which now occur, four are grasshoppers (Acrididae), and two of these, the field grasshopper (*Chorthippus brunneus*), and common green grasshopper (*Omocestus viridulus*) are both widespread and locally abundant. The remaining two, the meadow grasshopper (*Chorthippus parallelus*) and mottled grasshopper (*Myrmeleotettix maculatus*), are quite widespread but not so common. Only one bush-cricket (Tettigonidae), the oak bush-cricket (*Meconema thalassinum*), is still present, the dark bush-cricket (*Pholidoptera griseoaptera*) probably now being extinct.

The ground-hoppers (Tetrigidae) may resemble grasshopper nymphs but can be recognised by their greatly elongated pronota. Two of the three British species are now known to occur in Derbyshire. Unlike the grasshoppers and bush-crickets, which are adult during the summer and autumn, adult ground-hoppers may be found at any time of the year.

None of the three native species of cricket (Grylloidea) are found in the county, though there are old and dubious records of wood-cricket (*Nemobius sylvestris*), and mole-cricket (*Gryllotalpa gryllotalpa*). Two alien species occur however. The long established house-cricket (*Acheta domesticus*), which was formerly widespread in heated premises and on rubbish tips, has declined, due to the demise of coal for domestic heating and the introduction of insecticides, to the point where it has been recorded from few sites during recent years. A garden centre in north east Derbyshire is currently the only known British locality for the now cosmopolitan greenhouse camel-cricket (*Tachycines asynamorus*) whose original home was southern Asia.

The following species, included on early (pre 1905) county lists, are somewhat dubious, and doubt must be attached to their authenticity:

Great green bush-cricket	*Tettigonia viridissima*
Grey bush-cricket	*Platycleis albopunctata*
Bog bush-cricket	*Metrioptera brachyptera*
Wood-cricket	*Nemobius sylvestris*
Mole-cricket	*Gryllotalpa gryllotalpa*

Three of the seven native species, plus one which may now be extinct, have been recorded at ten or fewer sites in the county and are included here.

Nomenclature and arrangement follows Marshall and Haes (1988)

		National Status	*No. of post-1980 sites*
Oak bush-cricket	*Meconema thalassinum*		8

This pale-green insect is now the only known bush-cricket which still occurs in the county. It frequents Oak and other areas of broad-leaved woodland and has been found at a total of eight widely distributed sites in north-east, central and southern Derbyshire since 1972. Clearance of woodland or the replacement of broad-leaved trees by coniferous trees are the principal threats to its continued existence.

Dark bush-cricket *Pholidoptera griseoaptera* 0

In the list of species published by Jourdain in 1905, this insect was said to be rare at Repton Shrubs in southern Derbyshire. There have been no further records.

Slender ground-hopper *Tetrix subulata* 5

First found in the county in 1991 and now known from five sites. Two of these are oxbows in the Trent Valley; two are damp, industrial wastelands in the Erewash Valley and one is a small pond in N.E. Derbyshire. The latter is directly threatened by an opencast mining proposal.

Common ground-hopper *Tetrix undulata* 10

This small and inconspicuous insect has recently been recorded at ten sites, nine of them in or near the Peak District and the other at a Trent Valley power station. Five of the colonies are on rough grassland or scree slopes, sometimes associated with mineral waste, three are in disused quarries and two are on railway tracks. Three of the sites are Derbyshire Wildlife Trust reserves.

ACKNOWLEDGEMENT

I would like to thank Fred Harrison for providing copies of some of the older literature relevant to this article.

REFERENCES

FROST, R. A. (1991) The Status of Grasshoppers and Crickets in Derbyshire. *Derbyshire Entomological Society Journal,* autumn 1991, 12-21

JOURDAIN, F. C. R. (1905) The Orthoptera of Derbyshire. *Journal of the Derbyshire Archaeology and Natural History Society* 27, 229-232

MARSHALL, J. A. & HAES. E. C. M. (1988) *Grasshoppers and Allied Insects of Great Britain and Ireland.* Harley Books, Great Horkesley

WHITELEY, D. ed. (1985) *The Natural History of the Sheffield area and the Peak District.* Sorby Natural History Society, Sheffield

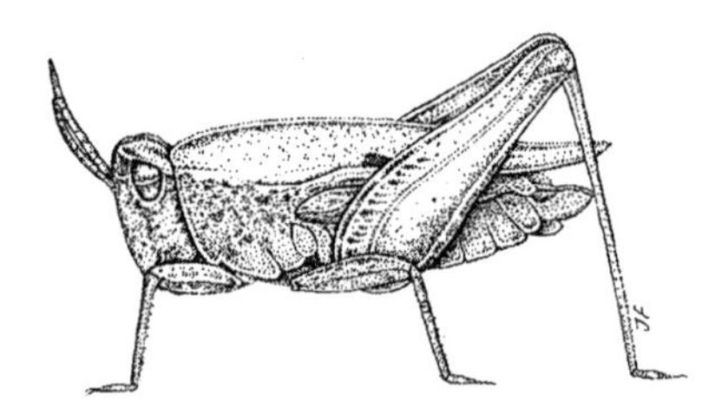

Common Ground-hopper

INDEX

English Names

Latin

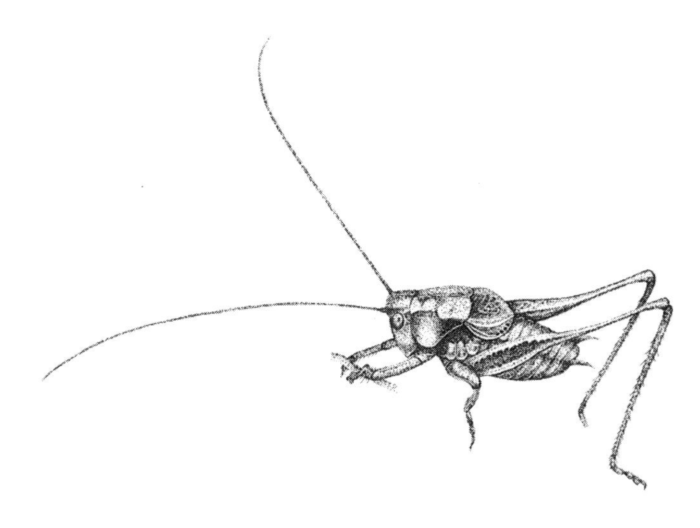

Dark Bush-cricket (male)

THE TRUE BUGS (Hemiptera - Heteroptera)

DAVID BUDWORTH

The Heteroptera are the smaller of the two suborders that make up the Hemiptera, with about 450 species in Britain. They have piercing/sucking mouthparts and feed on plant juices and/or animal body liquids. This group has not been popular with entomologists although interest in them has increased over the last decade. Some are easy to identify but other, more critical species need microscopic examination. Perhaps the most studied have been the aquatic families which have been surveyed along with other water dwelling invertebrates.

The national Red Data Book (Shirt, 1987), lists 14 endangered Heteroptera species, 6 vulnerable and 53 rare. Of the rare species, 11 have been only recently discovered in Britain. The Derbyshire records to date have not revealed any species within these categories. A more detailed review of the scarce and threatened Heteroptera within Britain has been published, (Kirby, 1992) and this provides a useful reference for species records from within the County.

The Heteroptera occupy all habitats other than marine and are most obvious to the casual observer during the months of July to September. In the earlier summer months many species are at immature stages and present yet further difficulty in identification.

Data on the County Distributions was published earlier this century, Bedwell (1945) and this listed 171 species for Derbyshire. An ongoing study of bugs throughout the County since 1975 has extended this list to over 220 species although some of Bedwell's species do not have any recent records. There are still large areas in the eastern Coal Measures and along the western border in particular, where more recording effort is needed. Short papers have appeared, listing the species records from specific areas, but a full assessment for the County is still in its infancy.

Making a judgement therefore of the significance and/or status of the bugs within the county can only be very tentative. Any short term dynamic variation in a given species may go unseen. With this inevitable lack of records it is still nevertheless possible to develop a knowledge of the range and distribution of the species over a longer period. Where this discloses creatures which are uncommon or rare, site management can be adjusted to maintain their presence.

The remainder of this section lists the Heteroptera species for which there is only one or two county records over the past 15 years and which are considered uncommon in the Midlands.

The nomenclature is according to Southwood & Leston (1959) but the current name is added in brackets where this has changed.

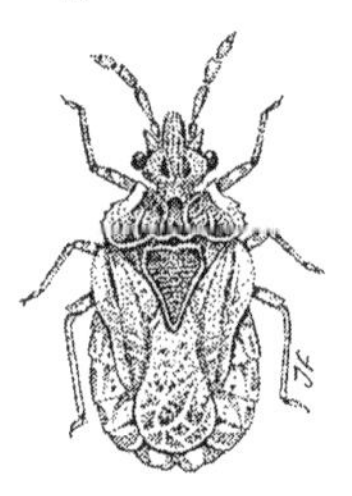

Aradus depressus

		National Status	No of post-1980 sites
Aradus depressus	Flatbug	Local	1

Only one record exists for Derbyshire and a second near the County border in NW Nottinghamshire. This species is described in the literature as being widespread and associated with freshly cut tree stumps such as silver birch and oak. Its local status must be considered unknown with such a paucity of records but the same situation exists in most other counties and a red data book category is probably inappropriate.

Elasmucha ferrugata Shieldbug Extinct 0

There are two records of this bug in the County. One from Derby in 1903 and one in Glossop, 1950. One of its foodplants is cowberry and as such it may well still be present, although its British status is now considered extinct.

Pitedia juniperina Shieldbug Extinct 0
(Chorochroa juniperina)

Southwood & Leston describe this species as an entomological mystery. There was a single record from Derbyshire in 1902 and the last record in Britain was from Lancashire in 1925. Now considered extinct in Britain but is a common species throughout Europe and North Africa.

Coriomeris denticulatus Squashbug Common in south 2

This species has a recorded range up to Yorkshire but is generally a southern species south of a line from Lincolnshire to Pembroke. The Derbyshire records are from Poolsbrook in the north to Ilkeston in the south. As a generally ground dwelling species it is easily overlooked and may well be under-recorded, being likely to be found on old spoil heaps, in quarries or demolition sites where black medick has colonised.

Drymus ryei Ground bug Local 1

The one record of this bug is from Monks Dale, Tideswell. It is described as widely distributed and to occupy a similar dry soil ground litter habitat as the much more common *D. sylvaticus*. The lack of records would suggest its possible requirement for a more specialised microhabitat but one would expect it to be present in the other Dales.

Drymus pilicornis Ground bug Notable B 1

There has been confusion over the identification of this species but genuine records are mostly from the chalk downs in the south. There is however a record from the Lathkill Dale.

Scolopostethus decoratus Ground bug Common in south 3

Southwood & Leston (1959) state that this bug can be easily found amidst heath and heather. The few records for Derbyshire, with its large areas of such vegetation, must imply that the insect is scarce. It is possible however that the species is missed by general sweeping and searching needs to be done at ground level.

		National Status	*No of post-1980 sites*
Cymus claviculus	Ground Bug	Notable in northern England	3

This species has been recorded mainly to the south of the County and is another ground dweller. It is interesting that two sites are power stations on the poor flash/gravel disturbed soils just becoming invaded with vegetation. The insect's habitat preference is for dry meadows (note that the Carvers Rocks, Hartshorne site is sandy heath) so it is possible that well drained soils on gravel are also suitable. The literature does however consider the Trent Valley to be its northern limit and as such it may well be limited to the south of the County.

Campylosteira verna	Lacebug	Notable in northern England	1

There is a record of this bug from Bakewell. The literature however defines its range as the chalk downs of southern England but for a doubtful record from Northumberland. It may well be that the Derbyshire distribution is limited to parts of the limestone area.

Amblytylus evanescens	Capsid bug	Notable B	0

This grassland species, listed by Bedwell (1945) as being present in Derbyshire, has not been recorded since. It is considered that the species' existence within Britain is dubious since it is easily confused with the southern species *A. nasutus*, which has also been recorded in Derbyshire prior to 1950.

Chlamydatus evanescens	Capsid bug	Category 3	0

This bug has only been recorded from the western counties of Cheshire, Denbigh and Caernarfon, but there is a single record from the Staffordshire side at Dovedale. It feeds on wall pepper and white stonecrop and searching the areas where this grows may well produce a record for Derbyshire.

Monosynamma sabulicola	Capsid bug	Notable B	1

This is a species of coastal dunes with occasional records inland at gravel pits and quarries. There is a single county record from Swarkestone Gravel Pits.

Halticus luteicollis	Capsid bug	Local in south	1

A single specimen alighted on the author's shirt in Newhall, Swadlincote. The bug's range is south of the line from the Wash to Glamorgan and the Trent Valley represents a northern limit. The host plants are the bedstraws and it can be found quite frequently on *Galium mollugo*. Further searching on this plant may well produce more site records.

Globiceps salicicola (Globiceps juniperi)	Capsid bug	Notable B	1

Confusion with this species and *D. cruciatus* may well have accounted for its inclusion in Bedwell's earlier list for the County, especially since it was not considered British until more recently. The single Derbyshire record was on Stanton Moor in 1978.

		National Status	*No of post-1980 sites*
Stenodema trispinosum	Grass bug	Local in south	1

Two or three specimens of this mirid bug were taken in 1992 on some marshy land near Poolsbrook, Staveley. The insect is associated with sedges, especially tufted sedge and has a south east distribution in Britain. Investigation needs to be made at other sites in the area to determine the size of this small population.

Salda muelleri	Shorebug	Notable in northern England	1

This bug, of wet margins of water bodies, has a discontinuous distribution with records from the Devon and East Anglia Fens as well as upland moorlands in Yorkshire, Lake District, Wales and Scotland. There is a recent record of this in Derbyshire from Matlock Moor.

Salda morio	Shorebug	Notable in northern England	1

This bug is recorded from Scottish and northern England counties and there is an early record from Buxton in 1889. The species has since been recorded on Edale Moor. It occurs in peat bogs at the margins of pools. The same situation has been noted as with the last species and its status must remain tenuous.

Ilyocoris cimicoides	Saucer Bug	Southern distribution	4

There are few records for this species in Derbyshire and these have been along the eastern area from Staveley to Ticknall. Southwood & Leston (1959) report its presence southwards from Nottinghamshire to Staffordshire but it is currently considered rare in Lincolnshire.

Glaenocorisa propinqua	Waterbug	Local	2

Although this is common in upland pools throughout Scotland and as far south as Cheshire it has been considered to be a retreating relic species with a single known locality remaining in the Somerset uplands. The species is now increasing in Scandinavia in the waters where fish are disappearing. There are two records in Derbyshire, one on Longstone Moor and the other on Featherbed Moss near the Snake Road, but it is expected that more sites could be verified by a thorough search.

Callicorixa wollastoni	Waterbug	Notable in northern England	4

Another upland species generally in localities above 300m. The Derbyshire Peaks are possibly the furthest south of its distribution in Britain. Recent records are from Leash Fen, Big Moor Reservoir, Longstone Moor and Featherbed Moss. It is again felt that more sites could be obtained by further searching.

ACKNOWLEDGEMENTS

I would like to acknowledge Mrs E. Thorpe & Mrs F. Jackson, Drs P. Kirby, A.C. Warne and D. Yalden. Messrs A. Brackenbury, W. Ely, F. Harrison, R. Merrit, A. Price, I. Viles, D. Whiteley and D. Young for their efforts in supplying Heteroptera records for the County.

I further wish to thank Dr P. Kirby for reading the draft version and providing further comments and additions to the various species accounts.

REFERENCES

BEDWELL, E. C. (1945) The County Distribution of the British Hemiptera-Heteroptera *Ent. Monthly Magazine 18.*

BUDWORTH, D. (1976) Heteroptera Survey. *Drakelow Wildlife Report.*

BUDWORTH, D. (1979) Heteroptera Survey. *Carvers Rocks Nature Reserve Survey.* A. Hopkins (private publication).

BUDWORTH, D. & KIRBY, P. (1978) Heteroptera of Derbyshire. A plea for records *Derbyshire Entomological Society Journal No. 53.*

KIRBY, P. (1992) *A Review of the Scarce & Threatened Hemiptera of Great Britain.* U.K. Nature Series No. 2. Joint Nature Conservation Committee, Peterborough.

MORRIS, M. G. (1975) *Preliminary observations on the effects of burning on the Hemiptera (Heteroptera and Auchenorhyncha) of limestone grassland.* Biological Conservation 7:311-319.

SAVAGE, A. A. (1989) *Adults of the British Aquatic Hemiptera-Heteroptera.* Freshwater Biological Association.

SHIRT, D. B. (1987) *British Red Data Book 2 - Insects.* Nature Conservancy Council, Peterborough.

SOUTHWOOD, T. R. E. & LESTON, D. (1959) *Land and Water Bugs of the British Isles.* Warne. London.

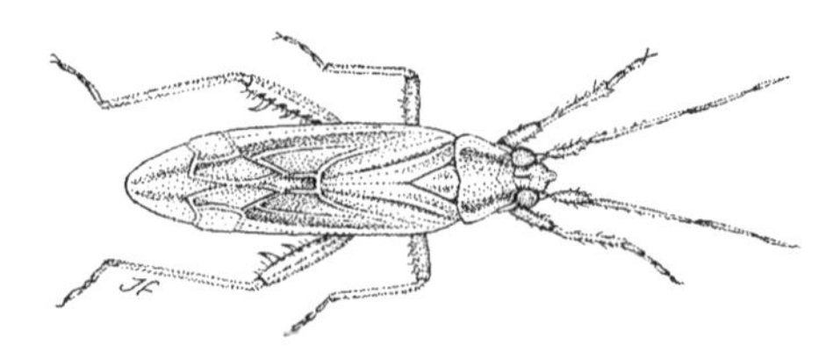

Stenodema trispinosum

INDEX

Scientific Names (Genera)

BEETLES (Coleoptera)

GRAHAM J. MAYNARD

Around 4,000 species of beetle have been identified so far in the British Isles and these occur in a very wide range of ecological niches. About 1500 beetle species have been recorded so far in Derbyshire.

Many of the historical records are imprecise and difficult to interpret. In order to present a concise and accurate picture from the mass of data the author has decided to restrict the scope of this account to the beetles of one important habitat.

The following list represents the rarest species which have been found in the relatively undisturbed parklands of Derbyshire. These records are taken from reports supplied by Colin Johnson of Manchester Museum and the original documents are held by the Derbyshire Wildlife Trust.

All the beetles listed are found in rotting or dying wood, or the associated tree fungi. Nomenclature and arrangement follows Kloet and Hincks, updated by Pope (1977).

The national status is taken from the RECORDER database supplied by JNCC. This is the only available complete check list which includes the national status reports for all beetles. The status has been cross referenced with Hyman revised by Parsons (1992, 1994). This revises Hyman (1986) and Shirt (1987), but does not include the status for beetles in the following groups:- Ptilidae, Staphylinidae, Pselaphidae, Cantharidae, Cryptophagidae, Lathridiidae for which information is taken from the JNCC database.

A survey of uncommon water beetles in the county has recently been published by Merritt (1995)

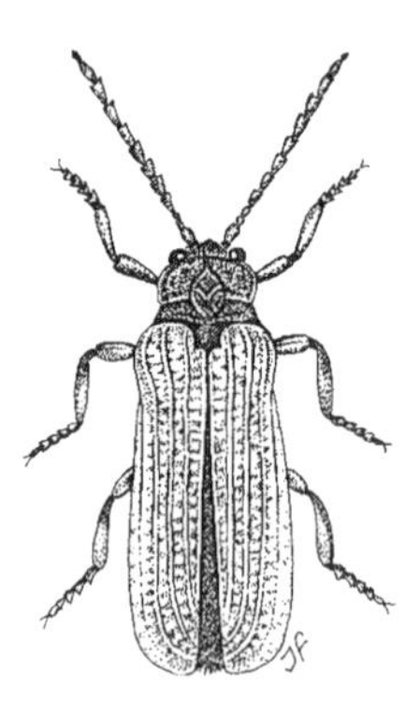

Pyropterus nigroruber

	National status	*No. of post-1980 sites*
Histeridae		
Plegaderus dissectus	Nb	2
Abraeus granulum	Na	1
Aeletes atomarius	RDB3	1
Ptilidae		
Nossidium pilosellum	Nb	1
Ptenidium gressneri	RDB3	1
Ptenidium turgidum	Na	1
Micridium halidaii	RDB1	1
Staphylinidae		
Omalium allardi	Na	1
Quedius microps	Nb	2
Quedius scitus	Nb	2
Sepedophilus testaceus	Nb	2
Gyrophaena angustata	Nb	2
Gyrophaena hanseni	Nb	1
Atheta basicornis	Nb	2
Phloeopora angustiformis	Nb	1
Pselaphidae		
Bibloporus minutus	Nb	1
Euplectus bonvouloiri rosae	Na	1
Euplectus fauveli	Nb	1
Euplectus kirbyi	Nb	1
Buprestidae		
Agrilus laticornis	Nb	1
Cantharidae		
Malthinus frontalis	Nb	2
Lycidae		
Pyropterus nigroruber	Na	1
Dermestidae		
Megatoma undata	Nb	2
Ctesias serra	Nb	2
Anobiidae		
Ptinomorphus imperialis	Nb	1
Dorcatoma serra	Na	1

	National status	*No. of post-1980 sites*
Trogossitidae		
Nemozoma elongatum	RDB3	1
Peltidae		
Thymalus limbatus	Nb	1
Cleridae		
Tillus elongatus	Nb	1
Korynetes caeruleus	Nb	1
Lymexylidae		
Hylecoetus dermestoides	Nb	2
Rhizophagidae		
Rhizophagus nitidulus	Nb	1
Cyanostolus aeneus	Na	1
Sphindidae		
Sphindus dubius	Nb	2
Cryptophagidae		
Cryptophagus labilis	Nb	1
Atomaria morio	pRDBk	1
Endomychidae		
Symbiotes latus	Nb	1
Lathridiidae		
Lathridius consimilis	Nb	1
Enicmus rugosus	Nb	1
Corticaria alleni	Nb	1
Mycetophagidae		
Mycetophagus piceus	Nb	1
Mycetophagus populi	Na	1
Tenebrionidae		
Prionychus ater	Nb	2
Mycetochara humeralis	Na	2
Melandryidae		
Melandrya caraboides	Nb	1
Conopalpus testaceus	Nb	2

	National status	*No. of post-1980 sites*
Oedemeridae		
Ischnomera caerulea	pRDB3	1
Aderidae		
Aderus oculatus	Nb	2
Cerambycidae		
Stenostola dubia	Nb	2
Curculionidae		
Cossonus parallelepipedus	Nb	1
Scolytidae		
Ernoporus caucasicus	RDB1	1

REFERENCES

HYMAN, P. S. (1986) *A national review of British Coleoptera. Part 1a. A review of the status of British Coleoptera.* Invertebrate Site Register, 64, 1-71. Nature Conservancy Council, Peterborough.

HYMAN, P. S. rev. by PARSONS, M. S. (1992) *A review of the scarce and threatened Coleoptera of Great Britain* (Part 1). JNCC (UK Nature Conservation No. 3)

HYMAN, P. S. rev. by Parsons, M. S. (1994) *A review of the scarce and threatened Coleoptera of Great Britain* (Part 2). JNCC (UK Nature Conservation No. 12)

MERRITT, R. (1995) Notable water beetles in Derbyshire: an interim review. *Sorby Record* 31, 59-64.

POPE, R. D. (1977) 2nd edition (completely revised) of Kloet, G. S. and Hincks, W. D. *A Check List of British Insects.* Part 3. *Coleoptera and Strepsiptera.* Handbooks for the identification of British Insects XI, part 3. Royal Entomological Society of London.

SHIRT, D. B. ed. (1987) *British Red Data Books: 2 Insects.* Nature Conservation Council, Peterborough.

HOVERFLIES (SYRPHIDAE)

EILEEN THORPE AND DEREK WHITELEY

Flies (Order Diptera) differ from other insects in that they have two wings instead of the more usual four, the back pair being modified into small balancing organs. They are a very large and important order; in Britain only the Order Hymenoptera, (which includes ants, bees and various types of wasp), has a greater number of species. Hoverflies (Syrphidae) are a distinctive and easily recognised family within this diverse order.

There are around 270 British species, and the known Derbyshire hoverfly fauna currently numbers 168 species. Hoverflies are attractive insects, conspicuous when they hover in mid-air or feed at flowers. Some are boldly coloured and patterned, some excellent mimics of wasps, honey bees or bumble bees; however hoverflies neither sting nor bite.

Adult hoverflies are important pollinators, especially of those flowers generally ignored by bumble bees. Nearly half the British species have predacious larvae, most feeding voraciously on aphids; others feed in living plant tissues, and the rest occur in a great number of different habitats both terrestrial and aquatic. There they are variously decomposers, detritus feeders and scavengers, contributing to the essential recycling of nutrients and organic wastes.

Hoverflies as a group have considerable potential as a conservation tool for site assessment, because of the variety of situations in which they can be found and the number of species with specialist life-styles, often of very local occurrence. Some species are strongly associated with ancient habitats with a history of consistent management. For instance there is a list of species indicative of ancient woodland, and among these, hoverflies breeding in dead wood can be good indicators of this important micro-habitat (Stubbs & Falk, 1987). Local dipterists have extended the concept of indicator species to assessing the quality of wetlands, using a combination of habitat fidelity and local scarcity, (Whiteley, 1987a).

NATIONAL STATUS

The British Red Data Book for Insects (Shirt, 1987) defines categories of threat and these are described in the 'Introduction to Arthropods', together with the lower rarity 'nationally notable' categories. *A Review of the scarce and threatened Diptera of Great Britain*, (Falk, 1991), revises the status of some hoverfly species in the Red Data Book and is the authority adopted here.

Great Britain is covered by a grid of more than 2800 10-km squares. Nationally scarce species are estimated to occur within the range of 16 to 100 of these, and in the case of hoverflies the records contributing to this assessment are those received since 1960. The 'nationally scarce' category provides a pool of species which is the main tool for evaluating sites for conservation purposes. Another important aspect of the national reviews for diptera and the other major invertebrate orders, is that they combine a status assessment for many species with detailed ecological information and advice on conservation management.

Since species are not equally scarce over the whole of their ranges, there are a number which although not scarce on a national basis are nevertheless important in regional faunas. Because of its geographical position and topography, Derbyshire holds faunas with both southern and north-western elements, and these are often of considerable scientific interest. In a European context the following recommendation is applicable to certain hoverflies, hymenoptera and many beetles. This is Recommendation R(88)10 of the Committee of Ministers of the Council of Europe, on the protection of saproxylic organisms and their biotopes (1988). It recognises the need to protect a whole endangered habitat, that of dead wood and old decaying trees, as one of the most threatened in Britain and over most of Europe (Fry & Lonsdale, 1991; Speight, 1989). It is discussed more fully in relation to hoverflies in the section 'Particular threats to the group'.

THE RECORD BASE

Lists of hoverflies were published for a few areas of Derbyshire in the early 20th century (Jourdain, 1905; Drabble, 1916 & 1918). These lists pose problems of identification in the light of at least two major taxonomic revisions of the family within the intervening period; also the localities of these early records were only vaguely defined. Consequently we have preferred to ignore them for the time being, and we have made no inferences from them regarding declines. This assessment is based on recent recording since 1975.

Derbyshire lagged behind Yorkshire in beginning the study of hoverflies, but an initiative taken by the Natural History Section of the Sheffield City Museum in 1971 subsequently grew in the mid-1970s to involve members of the Sorby Natural History Society of Sheffield, with a recording area covering north Derbyshire and south Yorkshire. With support from the Museum and Sheffield University, Derek Whiteley formed the Sorby Diptera Group which has provided the mainstream interest and focus for recording hoverflies in this area. A further upsurge of recording activity followed the publication of the hoverfly key (Stubbs & Falk, 1983). Hoverflies have proved a popular group among amateur entomologists and the interest has been well sustained up to the present. A review of the hoverflies of the Sheffield area and north Derbyshire was published by the Sorby Society in 1987, based on approximately 8000 records (Whiteley, ed. 1987b). The national Hoverfly Recording Scheme was launched in 1976 and a close relationship developed with the national organisers who provide an immense amount of support to local recorders. This scheme has been well supported, and the current national data base stands at well over 200,000 records.

Limitations of recording

Because the main recording initiative stemmed from the Sorby Society's Diptera group, the hoverfly fauna of north and mid-Derbyshire is much better known than that of the south, and coverage over the county as a whole has been very uneven. This imbalance remains despite increasing activity from members of the Derbyshire Entomological Society from the mid-1980s, and from three years of intensive survey in the south of the county by the authors from 1989 to 1991. An attempt has been made to include all records up to and including the 1994 season to improve the data base for south Derbyshire. It is, however, fair to comment that good habitat is more highly fragmented in the south of Derbyshire than elsewhere in the county. Decisions to omit some species considered most likely to be under-recorded have been influenced to some extent by comparisons with the number of south Yorkshire localities, although such decisions must remain provisional until the ecologically distinct region of south Derbyshire has been more fully sampled.

Sources of information

By far the greater part of the information was extracted from The Sorby Record Special Series No.6. (Whiteley, D. ed. 1987b). This was updated from the files of the Sorby Natural History Society held in Sheffield City Museum, and in private collections, to include records up to 1994. Additional information has come from: DES Quarterly Journals, numbers 91, 99, 100, 104, 108, 110, 111 & 112 (1988-1993); The Entomologist's Record, March and June 1958; National Trust surveys of its Derbyshire properties from 1984; individual members of the Derbyshire Entomological Society from 1984; Derbyshire lists from W. Ely of Rotherham Museum, 1979-1985; Derbyshire lists from B.Wetton of Nottingham, 1987-1994.

CRITERIA FOR INCLUSION

1. The list shall include all species of national Red Data Book status as revised by Falk (1991).

2. Species known from fewer than 10 localities in the county. Unless otherwise indicated this number refers to post-1980 records. Exceptions are the few cases where records exist only from the years 1977-1979 for sites which it has not been possible to re-survey. These sites have been included in the present assessment and the date of the latest record is indicated in the text.

3. Some species within this category of fewer than 10 sites have been further assessed by using the following qualifying habitat criteria: species dependent on particularly scarce and fragmented habitats, and on habitats particularly prone to destruction. Such habitats of importance to hoverflies are ancient woodland with dead wood and large, old, decaying trees; a number of scarce types of wetland including some that are small and often overlooked; unimproved grassland and moorland especially with vegetation mosaics. An additional requirement to all these habitats is the associated presence of flower-rich habitat for adult feeding.

The habitat descriptions given in the text of the systematic list are intended to be taken into account in addition to the number of localities, when assessing the Derbyshire status of a species. All descriptions have been abbreviated from a fuller text which will be kept by the Derbyshire Wildlife Trust.

PARTICULAR THREATS TO THE GROUP

Hoverflies are threatened by several aspects of site management, including management where conservation is an objective, but where the special needs of insects are not given due consideration. The concern is loss of specialist niches in habitats, often where those habitats remain ostensibly suitable. Losses can be caused both by disruptive or by gradual habitat changes induced by inconsistency and changes in management.

An example in woodlands are the hoverflies breeding in dead wood and, more usefully, in decaying parts of ancient, living trees. The invertebrate fauna of this habitat involves some 1000 species, with flies and beetles predominating. Dead wood is not a uniform habitat; there are whole successions of communities that are often interdependent and they collectively exploit many different types of niche. Some of their more specialist niches are extremely rare and include particular states of internal rot that may not arise till living trees are upwards of 200 years old. Hoverflies are valuable for indicating the survival of such important microsites in a woodland context.

Dead-wood hoverflies are threatened by removal of dead timber and over-mature decaying trees; by excessive tree surgery and 'tidying up'; and by failure to maintain a varied age structure of trees in woodland. Recently it has become fashionable to introduce or resume coppice management in some neglected woodlands and this can also be damaging if done without adequate preliminary site assessment, (Watkins, 1990). There are about 46 British dead-wood hoverflies; 25 are recorded for Derbyshire but sites with a rich dead-wood hoverfly fauna are few, and the number of such species listed here reflects the highly fragmented nature of suitable habitat remaining in the county.

Practically all hoverflies breeding in woodland are also threatened by failure to maintain both adult and larval feeding habitats sufficiently close together. Adults need nectar and pollen to fuel their highly energetic life styles, and to cater for reproductive demands. This means an abundance, succession and variety of flowers in close proximity to the breeding habitat. Adult feeding sites in woodlands are threatened by shading out of rides and sunny clearings, or by lack of a natural wood-edge of flowering shrubs and flowery margins. Regrettably, this habitat is seldom created in many new and amenity tree plantations. Consequently many present-day woodlands are quite inadequate, not only for hoverflies but for all insects with similar adult requirements.

Among wetland habitats, those of special importance for flies with aquatic or semi-aquatic larvae are often fringe habitats, shallow water areas and marshes with bare mud; sometimes quite small features such as flushes and seepages that are very often overlooked and under-valued. Margins of open water such as gravel pit ponds and reservoirs can be of variable quality for hoverflies; the need is for shallow bank profiles with a well-zoned marginal vegetation. Wetland hoverflies are threatened by land drainage, by lowered ground water levels leading to scrub invasion and drying out; by enrichment of water inflows largely from agricultural run-off; and by extensive clearance of the natural vegetation of water margins, and ditches. Whilst clearance is often necessary to maintain the habitat, a rotational regime should always be adopted, with at least some late stages in the succession present every year. Good habitats are also frequently threatened by excessive trampling of water margins in heavily fished waters.

In grassland communities some hoverflies are threatened by natural succession to scrub and coarse grasses with loss of short, flowery turf and bare ground mosaics. This ecological grouping includes many insects from other orders such as the solitary bees, and wasps, and members of other fly families. Regularly disturbed or early successional habitat is in Derbyshire largely concentrated in disused industrial sites with the better ones retaining some pockets of older habitat. Such sites have a high rate of loss to re-development, but even in those which escape this threat the habitat is at risk unless the site is suitably managed.

'ALMOST RARE' SPECIES IN DERBYSHIRE

In this section is a short list of species which do not meet the criteria for inclusion in the main list, but which it is thought deserve mention:

1. Species whose populations need monitoring as they require specific habitats which have a high rate of loss, or are somewhat ephemeral:

Chalcosyrphus nemorum, 11 sites

Brachypalpoides lenta, 10 sites, both breeding in dead/decaying wood. These species may have temporarily elevated populations because of the widespread effects of Dutch Elm disease on wych elms in our local woods.

Paragus haemorrhous, 13 sites, many on derelict industrial land which is scheduled for development, or is currently lacking essential management to maintain the habitat.

2. Nationally scarce species whose habitats are apparently particularly well represented in Derbyshire, and for which we have a special responsibility:

Neoascia obliqua (N), 12 sites, associated with riverine marshes, especially along the middle reaches of the Derwent, the Wye and the Moss.

Didea fasciata (N), 10 sites, associated with old woodland.

ENDANGERED DERBYSHIRE HOVERFLIES

Nomenclature and Systematic order follow Stubbs & Falk (1983).

National status is taken from Falk (1991).

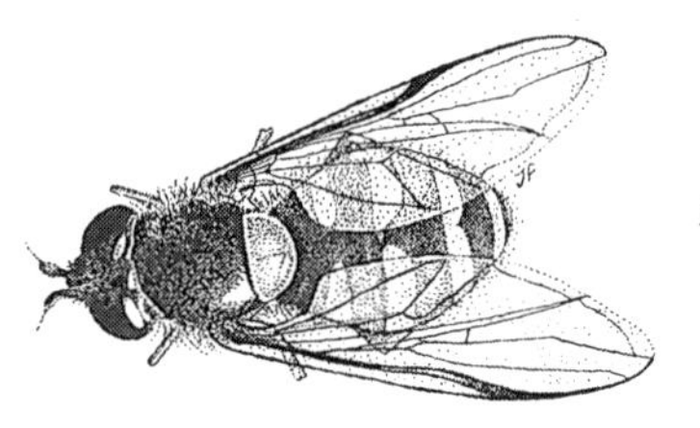

Megasyrphus annulipes

	National status	*No. sites in Derbyshire*
Platycheirus ambiguus		2

Adults in sunny, sheltered wood margins with early flowering blossom.

Platycheirus discimanus	N	1

Broadleaved woodland, with open sunny rides and clearings, or scrub areas of heath.

Platycheirus perpallidus	N	1

Marshes, especially with sedges, in wet, poor fen or boggy areas at the margins of ponds, lakes and rivers.

	National status	*No. sites in Derbyshire*
Platycheirus podagratus	N	1

Bogs, poor fen and river margins in both lowland and sub-montane areas exceeding 450m. Acidic conditions may be a requirement

Xanthandrus comtus	N	2

Flowery woodland rides and margins, and scrubland.

Chrysotoxum festivum		1

Typically in grassy places, usually near woods or scrub margins. A second site has been lost since 1985.

Didea intermedia	N	2

Habitat uncertain, but apparently associated with some coniferous woods, or heathland with adjacent flowery areas. In the former case there is a reliance at the Derbyshire site on adjacent roadside verges for flowers.

Epistrophe diaphana	N	1

Last recorded in 1979. A southern species of broadleaved woodland, especially margins and adjacent unimproved meadows.

Epistrophe nitidicollis		3

Open, sunny, woodland rides and flowery coppice glades.

Epistrophella euchroma	RDB3	1

Broadleaved woodland with open, sunny rides, clearings and wood edges. There is an association with old woods.

Megasyrphus annulipes	N	2

Often found in coppiced woodland, and at woodland edges and in rides.

Melangyna quadrimaculata		2

Deciduous broadleaved woodland with early flowering shrubs on sunny margins.

Melangyna guttata	N	8

Broadleaved woodland with flower-rich rides and clearings and wood margins. Old trees are a probable requirement.

Melangyna triangulifera	N	2

Broadleaved woodland, with flowery rides, clearings or margins.

Metasyrphus latilunulatus	N	3

Known from woods and heaths, preferences unclear.

Known from woods and heaths, preferences unclear.

	National status	*No. sites in Derbyshire*
Parasyrphus lineolus		1

Woodland, possibly adapting to coniferisation.

Parasyrphus malinellus		1

Flowery rides in coniferous woods and plantations.

Sphaerophoria batava		1

Flowery woodland rides, or wood edges associated with heathland.

Sphaerophoria rueppellii		3

Dry, unimproved and flower-rich grasslands in open sunny positions.

Xanthogramma citrofasciatum		3

Dry grasslands, especially where the turf is short and some bare ground is exposed, as along paths. Larvae have been found breeding in nests of the black meadow ant.

Callicera aenea	RDB3	1

Last recorded in 1977. Broadleaved woodland and old trees with cavities and rot holes.

Cheilosia antiqua		2

Last recorded in 1979. Broadleaved woodland, often in coppices with primroses, the larval foodplant.

Cheilosia carbonaria	N	2

Broadleaved woodland with flowery rides and margins.

Cheilosia cynocephala	N	1

Flowery grassland, usually on calcareous soils where the foodplant, musk thistle is abundant.

Cheilosia honesta		4

A predominantly southern species of deciduous broadleaved woodland, especially floristically rich wood margins.

Cheilosia intonsa		2

Flowery grasslands, including floodplain grassland, and sites adjacent to wet areas.

Cheilosia mutabilis	N	4

	National status	*No. sites in Derbyshire*
Chelosia praecox		3

Wood margins or areas associated with moist woods with early flowering shrubs.

Cheilosia pubera	N	3

Fens, fen carr, wet pastures, and water-margin marshes, possibly requiring base-rich conditions. In Derbyshire all sites are in the limestone dales.

Ferdinandea ruficornis	N	1

Old trees with persistent sap-runs as those infested with goat moth, in broadleaved woodland with a varied age-structure.

Brachyopa scutellaris		3

Old broadleaved woodland; mature and over-mature trees with sap runs.

Chrysogaster chalybeata		8

A wetland species of lush flowery meadows and fens, often associated with wood margins or scrub.

Chrysogaster virescens		3

A wetland species of bogs and wet acid pastures.

Neoascia geniculata	N	4

Marshes and water margins where there is lush emergent vegetation such as sweet reed-grass.

Neoascia interrupta	N	1

Lush ditches, ponds, fens and marshes associated with reed-mace and flower-rich marginal vegetation.

Orthonevra brevicornis	N	2

Marshes and fens, especially where seepages and streamlets are present.

Sphegina kimakowiczi		7

Dead wood in wet, shady areas or water margins of broadleaved woodlands.

Sphegina verecunda	N	3

Dead wood in shaded areas near streams in broadleaved woodland.

Anasimyia contracta		6

A wetland species associated especially with reed-mace at the margins of ditches, ponds and lakes.

	National status	*No. sites in Derbyshire*
Anasimyia transfuga		2

A wetland species of water margins with tall emergent reed-swamp vegetation, such as stands of common burr-reed.

Eristalis abusivus		5

Marshland habitat and mud at the edge of ponds.

Eristalis rupium	N	7

Upland cloughs near wet flushes and plenty of flowers, both in sheltered valley bottoms and glades, or more rarely in exposed situations up to 300m.

Helophilus trivittatus		4

A wetland species usually in grassy ponds in meadows, and in ditches.

Mallota cimbiciformis	N	1

Broadleaved woodland and parkland with a requirement for old trees with rot-holes.

Parhelophilus frutetorum		3

A southern species of open pond sides and ditches, often with reed-mace.

Parhelophilus versicolor		3

Open ponds and ditches, and strongly associated with reed-mace.

Eumerus strigatus		2

Asssociated with bulbous plants in woodlands, and also occasionally in gardens. Mysteriously uncommon in Derbyshire.

Psilota anthracina	RDB2	1

Deciduous broadleaved woodland and parkland, almost exclusively in ancient sites.

Heringia heringi		2

Wood margins, specialising on wax-secreting aphids for prey, as in tree galls.

Pipiza bimaculata		4

Wood margins, specialising on wax-secreting aphids for prey.

Trichopsomyia flavitarsis		2

A northern species of wet heath, grasslands and wood margins.

	National status	*No. sites in Derbyshire*
Triglyphus primus	N	2

A wide range of habitats including woodlands and rough grasslands with scrub, specialising on predation of wax-secreting aphids, probably on roots, but the host-plant(s) is uncertain.

Arctophila fulva		2

A northern and western species of damp woodland margins.

Brachypalpus laphriformis	N	2

Large, dead and rotting tree stumps, especially of oak, in ancient broadleaved woodland.

Criorhina asilica	N	1

Last recorded in 1979. Dead wood habitat in well-structured woodland and old trees.

Criorhina floccosa		5

Dead and rotting wood in old, usually large, decaying trees in old woodland.

Criorhina ranunculi	N	5

Dead and rotting wood in old decaying trees, usually in old woodland.

Xylota coeruleiventris	N	1

Dead wood habitat; adults in this site occur over a wide area on the margins of a single large conifer plantation, coming to roadside flowers. The species is thought to be colonising some plantations from an original habitat in the Caledonian Pine Forest.

Xylota florum	N	2

A southern species, using dead wood in moist wooded valleys or woods with water.

Xylota tarda	N	1

Old decaying trees in wooded parkland.

Xylota xanthocnema	N	1

Dead hollow tree stumps and rotting cavities in old woodland.

ACKNOWLEDGEMENTS

1. The bulk of the recording was done by field entomologists in the Sorby Natural History Society of Sheffield, the records held partly in files at the Sheffield City Natural History Museum and partly in private collections.

2. Sheffield City Natural History Museum, for access to their files, and to their specimen collections. Also for copies of the lists of P. Skidmore (1953, 1955) and W. Dean (1975-1979).

3. F. Harrison for information from the files of the Derbyshire Entomological Society; also for copies of old county lists published in 1905, 1916 and 1918.

4. Dr A. Warne and other members of the Derbyshire Entomological Society for collection of specimens.

5. Dr K. Alexander and D. Clements for information from the National Trust Surveys.

6. W. Ely of Rotherham Museum for Derbyshire records.

7. B. Wetton of Nottingham for Derbyshire records.

8. For detailed information concerning ecology, habitat, specific threats, conservation and management objectives we have referred extensively to Steven Falk's 1991 *Review*; also to *British Hoverflies*, and to volumes of *Dipterists Digest*. These sources of information have been used for the species' accounts in the copy to be retained by the Derbyshire Wildlife Trust.

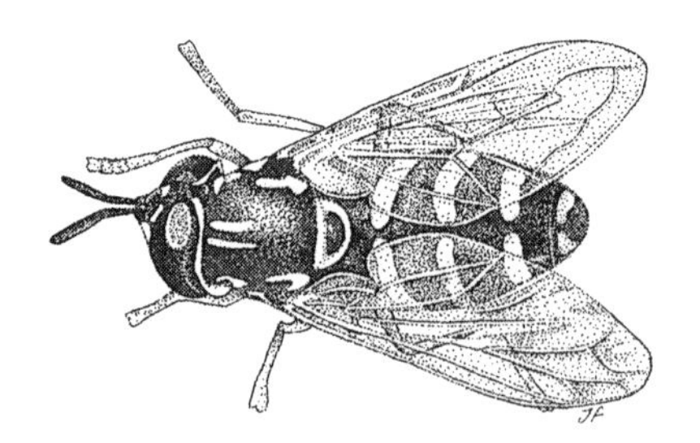

Chrysotoxum festivum

REFERENCES

DRABBLE, E. & H. (1916 & 1918). Notes on the Diptera of Derbyshire. *The Entomologist* Vols. 49 & 51.

FALK, S. (1991). *A Review of the Scarce and Threatened Flies of Great Britain (Part 1)* Nature Conservancy Council, Peterborough. (Research and Survey in Nature Conservation. No. 39).

FRY, R. & LONSDALE, D. eds. (1991). *Habitat Conservation for Insects – a neglected green issue.* The Amateur Entomologists' Society.

JOURDAIN, F. R. C. (1905). *Diptera.* in Page. W. (ed.). *Victoria History of the County of Derby.* Vol. 1. Constable, London. pp. 94-99.

SHIRT, D. B. ed. (1987). *British Red Data Books: 2. Insects.* Nature Conservancy Council Peterborough.

SPEIGHT, M. (1989). *Saproxylic invertebrates and their conservation.* Council of Europe. (Nature and Environment series, No. 42).

STUBBS, A. E. & FALK, S. (1983). *British Hoverflies.* British Entomological & Natural History Society.

STUBBS, A. E. & FALK, S. (1987). Hoverflies as Indicator Species in Whiteley, D. ed. (1987b). *Hoverflies of the Sheffield Area and North Derbyshire.* Sorby Special Series No. 6. Sorby Natural History Society, Sheffield City Museum & Derbyshire Wildlife Trust. pp. 46-48.

WATKINS, C. (1990). *Woodland Management and Conservation.* David & Charles for the Nature Conservancy Council, Peterborough.

WHITELEY, D. (1987a). Wetland Hoverflies in Whiteley, D. ed. *Hoverflies of the Sheffield Area and North Derbyshire.* Sorby Record Special Series No. 6. Sorby Natural History Society, Sheffield City Museum & Derbyshire Wildlife Trust. pp. 49-50.

WHITELEY, D. ed. (1987b). *Hoverflies of the Sheffield Area and North Derbyshire.* Sorby Record Special Series No. 6. Sorby Natural History Society, Sheffield City Museum & Derbyshire Wildlife Trust.

WHITELEY, D. ed. (1988-1994). *Dipterists Digest.* Vols. 1 to 14.

INDEX
Scientific Names (Genera)

BUTTERFLIES

Fred Harrison and Roy Frost

Since the earliest records of Derbyshire Lepidoptera were published in 1828, 48 out of a possible 68 or so species on the British Butterfly List have been observed at one time or another in the county. Of the recorded species, 38 are, or have been, present in the county during recent years and 9 are considered to be extinct. One migrant species, the Pale Clouded Yellow, has not been recorded for several decades and the occasional sighting of Swallowtail, White Admiral, Marbled White and Monarch are believed to be due to accidental or deliberate introduction of these butterflies.

At present only 17 of the resident species are relatively well established (although the degree of abundance and distribution varies considerably between individual species) and the remaining 13 are now so scarce and/or localised with regards to the distribution of their colonies, that they are considered to be endangered and thus qualify for inclusion in the Red Data Book.

The fortunes of these insects can undergo rapid change in a short space of time: the decline of the White-letter Hairstreak is associated with the destruction of elm trees due to Dutch Elm Disease from the mid-1970s onwards, but the reasons for the decline of other species are often not so obvious. Both the Holly Blue and Comma were for many years considered to be rare species in Derbyshire, but their numbers increased from the mid-1970s onwards and by 1992 both species were common and widely distributed. Populations of these butterflies suffered a rapid decline the following year however, and both are now considered to be scarce enough to qualify for Red Data Book status.

None of Derbyshire's resident butterflies qualify for inclusion in the National Red Data Book, but two are Nationally Scarce Species (Na) — Wood White, and White-letter Hairstreak, and six (Green Hairstreak, Brown Argus, Small Blue, Small Pearl-bordered Fritillary, Dark Green Fritillary, and Silver-washed Fritillary), are species that have suffered substantial local declines. They are included in the Nature Conservancy's (now English Nature) list of 15 species which merit attention in those regions (in our case the East Midlands) where declines have taken place.

The Green Hairstreak is now probably extinct in lowland areas of the county, but is locally common on the northern moorlands, whilst the Small Heath too is now predominantly an insect of upland areas; the latter butterfly's lowland population is virtually restricted to poor quality unmanaged grasslands on derelict industrial sites, disused railway lines, colliery spoil heaps, etc.

Thirty years ago the Small Skipper, Purple Hairstreak, Speckled Wood, Hedge Brown and Ringlet were considered to be extinct in Derbyshire. Since then all have re-colonised the county and continue, despite the fluctuating fortunes of some other species, to expand their range. The Small Skipper has become so well established that it is often the most abundant butterfly at many sites during July and August. The Purple Hairstreak's increase has been most noticeable during the last few years, but although one or two large colonies are known to exist in southern Derbyshire, it is too early yet to feel confident that it will be able to maintain its more northerly colonies in the long term. The Speckled Wood has increased its range considerably from a few small colonies located in the extreme south of the county to the extent that it is now locally common throughout much of southern Derbyshire with wanderers beginning to find their way into the Peak District dales. From its initial base in south-east Derbyshire, the Hedge Brown has expanded its range considerably and is now widely distributed throughout the southern half of the county and along the Magnesian limestone belt in the north-east. During the last few years it has begun to establish colonies in the less hospitable Coal Measures region of the county and has also reached the southern fringe of the Peak District. The Ringlet has been the last of our local butterflies to undergo a population expansion and appears to be following similar migration routes to the Hedge Brown, though at a slower pace.

All of the above species had been present in Derbyshire during the nineteenth century and had subsequently declined to the extent that they were considered to be extinct at the beginning of the present century. It remains to be seen whether or not the current increase of local populations of the previous five species is ephemeral, as appears to be the case with the Holly Blue and Comma, or represents the initial stages of long-term colonisation by these attractive insects.

Butterflies are sun-loving, day-flying insects, which depend to a great degree upon favourable weather conditions predominating during critical periods of their life cycle, especially during mating flights and egg laying. If prolonged periods of cold wet weather persist at these times, then the ability of the adults to produce fertile eggs for the next generation is severely curtailed, and the population can suffer a drastic reduction. A poor spring with late frosts persisting into May or even June will be highly detrimental to those species which hibernate as adult insects, for they must find sustenance and mates after awakening from their long winter rest, and if they are unable to do so, population losses can be so severe that it takes them several years to recover. Most species can recover quickly from the effects of a poor breeding season, but several consecutive ones can be disastrous, and those whose colonies are small, isolated, and at the limit of their range may be driven into extinction.

Nationally the greatest variety and density of butterfly populations occur south of a line drawn between the Severn Estuary and the Wash, and the location of Derbyshire north of this means that many local species are at the limit of their range, particularly in those areas where the lowlands merge into the higher ground of the southern Pennines. In this situation, butterflies are particularly susceptible to periodic and sometimes extreme fluctuations in their population densities, but during the last twenty years they have generally been enjoying a favourable trend, possibly as a result of ameliorating climatic conditions.

Whilst climate has undoubtedly the greatest influence upon our local butterflies, other environmental factors too exert a great influence, and unless there is an adequate supply of suitable habitats, no amount of fine weather is likely to boost their population densities or save a species whose particular type of habitat is destroyed.

On the positive side, there has been a general reduction in the intensity of landscape management practices on non-agricultural land, e.g. weed spraying and grass mowing, with reduced cultivation of marginal land, all of which has undoubtedly been of benefit to wildlife. The creation of motorway systems with their considerable areas of grassy verges, and the conversion of miles of disused railway to amenity footpath routes, have created a network of wildlife corridors which enable species to disperse more easily — there is evidence for this on distribution maps showing the spread of well-documented species such as the Speckled Wood and Hedge Brown.

On the debit side, those habitats which support the greatest variety and density of species, the limestone grasslands, woodland rides and clearings, and long derelict industrial wasteland, have all suffered significant losses. This is a result of the lack of effective management to maintain a mosaic of vegetation types, including short-turfed, flower-rich, areas of grassland. The sparse, low nutrient 'soils' on many areas of derelict land enabled very rich communities of plants and insects to flourish without any effective form of management for long periods, but if undisturbed, these sites too will lose their diversity as a result of scrub encroachment, and now many have also been lost to re-development.

Those species of butterfly which are in the greatest danger of extinction may be dependent upon a particular foodplant (which itself may be scarce), may have very precise habitat requirements, and have a poor dispersal capability, and once the local colonies are lost, they may never be able to return to the county. Ten species are already listed as extinct and it is to be hoped that none of the 13 endangered species listed below will be added to them.

TABLE 1. Migrant, Vagrant and Extinct Species of Derbyshire Butterflies

The Swallowtail	extinct/introduced
Pale Clouded Yellow	migrant
Clouded Yellow	migrant
Black-veined White	extinct
Duke of Burgundy Fritillary	extinct
White Admiral	vagrant/introduced
Purple Emperor	extinct
Red Admiral	migrant
Painted Lady	migrant
Large Tortoiseshell	extinct
Camberwell Beauty	migrant
Pearl-bordered Fritillary	extinct
High Brown Fritillary	extinct
Marsh Fritillary	extinct
Marbled White	introduced
The Grayling	extinct
Large Heath	extinct
The Milkweed (or Monarch)	migrant/introduced

Small Pearl-bordered Fritillary Butterfly

Nomenclature follows Bradley and Fletcher (1986)
National Status: Hadley (1984), Shirt (1987)

		National Status	*No. of Post-1980 1 Km. Squares*
Grizzled Skipper	*Pyrgus malvae*		1

Occasionally present in small ephemeral colonies: also a vagrant. Poor quality, ungrazed grassland, in sunny sheltered locations. Threatened by inclement climatic conditions and scrub encroachment.

Wood White — *Leptidia sinapis* — Na — 1

One small colony temporarily established and possible vagrant. Woodland edges, rides and clearings in lowland areas. Threatened by inclement climatic conditions and scrub encroachment.

The Brimstone — *Gonepteryx rhamni* — 10+

Widely distributed but generally uncommon. Woodland edges, rides, and areas of scrub where the larval foodplant, buckthorn, grows. Threatened by inclement climatic conditions and low frequency of foodplants.

Purple Hairstreak — *Quercusia quercus* — 10+

Scattered colonies in south and east of the county. Edges, rides and clearings of oak woodland. Threatened by inclement climatic conditions and loss of mature oak woodland.

White-letter Hairstreak — *Strymonidia w-album* — Na — 10+

Rare and locally distributed. Woodland edges, rides, and clearings; parkland and hedgerows, wherever mature elm trees are present. Threatened by loss of mature, seed-bearing elm trees.

Small Blue — *Cupido minimus* — 3

Rare and locally distributed. Sheltered, south-facing, ungrazed sites with scattered scrub on the Carboniferous limestone. Threatened by loss of larval foodplant (kidney vetch) through the development of coarse herbage and scrub.

Brown Argus — *Aricia agestis* — 10+

Locally distributed at sites on the Carboniferous limestone. Warm south facing slopes with a mosaic of short turf and scattered scrub. Threatened by loss of larval foodplant (Rock rose) through the development of coarse herbage and scrub.

Holly Blue — *Celastrina argiolus ssp. britanna* — 10+

Widely distributed in small colonies, parks, gardens, hedges and woodland edges where holly and ivy are present. The status and distribution of this species are subject to severe fluctuations periodically which may be due to a combination of climatic and/or parasitic population increase factors.

		National Status	No. of Post-1980 1 Km. Squares

The Comma *Polygonia c-album* 10+

Widely distributed but generally uncommon. Woodland rides, parks and large gardens. Has a history of severe fluctuations in status and distribution nationally and only common in Derbyshire during recent years but now in decline again due to unknown factors.

Small Pearl-bordered Fritillary *Boloria selene selene* 2

Rare and locally distributed. Sunny sheltered rides and clearings amongst woodland or scrub. Threatened by loss of open areas through scrub encroachment.

Dark Green Fritillary *Argynnis aglaja aglaja* 10+

Small colonies, widely distributed, principally on the Carboniferous limestone. Sheltered, south-facing rides and clearings amongst woodland or scrub. Threatened by loss of open areas through scrub encroachment.

Silver-washed Fritillary *Argynnis paphia* 7

Rare and of sporadic appearance at widely scattered locations; possibly a vagrant. Woodland rides and clearings. Threatened by inclement climatic conditions and loss of open areas through scrub encroachment.

The Ringlet *Aphantopos hyperantus* 10+

Uncommon and at present restricted to lowland areas in the south and east of the county. Sunny, sheltered woodland edges, rides and clearings. Threatened by inclement climatic conditions and loss of open areas in its woodland habitat.

ACKNOWLEDGEMENTS

We are grateful for the assistance and constructive comments during preparation of the text by Philip Bowler and Ken Orpe.

REFERENCES

BRADLEY, J. D. and FLETCHER, D. S. (1986) *An Indexed List of British Butterflies and Moths*. Kedleston Press, Orpington.

HADLEY, M. (1984) Invertebrate Site Register, Report No. 46, Nature Conservancy Council, Peterborough.

HARRISON F. & STERLING M. J. (1985) *Butterflies and Moths of Derbyshire Part 1*, Derbyshire Entomological Society, Derby.

HEATH J. & EMMET A. M. (1989) *The Moths and Butterflies of Great Britain and Ireland.* Vol. 7. Harley Books, Colchester.

HEATH J., POLLARD E., & THOMAS J. A. (1984). *Atlas of Butterflies of Great Britain and Ireland.* Viking Publications, Harmondsworth.

Nature Conservancy Council (1985), Peterborough. *The Management of Chalk Downlands for Butterflies*. Focus on Nature Conservation Series No. 17. Peterborough.

SHIRT, D. B. (1987) *The British Red Data Books: 2 Insects.* Nature Conservancy Council, Peterbough.

Small Blue Butterfly

INDEX

English Names

Scientific Names (Genera)

THE LARGER MOTHS (MACRO-LEPIDOPTERA)

FRED HARRISON AND IAN VILES

The total number of larger moths on the Derbyshire List currently stands at 540, from which may be deducted 15 migrants from the continent who cannot survive the rigours of our winters, and another 39 species (see Table 1) that are native to Britain, but due to their rarity and sporadic appearance are considered to be vagrants so far as Derbyshire is concerned. This leaves 488 species of moths that are regarded as residents, of which 31 (6.35%) have not been observed during recent times and are considered to be extinct (see Table 2). Of the remaining residents, 122 (25%) have been recorded in such low numbers, or are so restricted with regard to the number of their surviving colonies, that they are considered to be endangered species. Only one of these is of sufficient scarcity in Britain as a whole to qualify for inclusion on the national list of Red Data Book species, but 17 are on the Annotated List of rarer British macro-moths (from Waring, 1994) graded either Na or Nb, and a further 59 species are classed as nationally local.

Derbyshire is a particularly interesting county in which to study lepidoptera; the southern tip of the Pennine Hills extends southwards almost to the Trent Valley, and the complex geology, topography, and micro-climates associated with the uplands and valleys create both a barrier to the northward expansion of many southern lowland species, and enables many northern upland species to survive at the extreme southern limit of their range. Southern species do occur north of Derbyshire in the lowland areas both east and west of the Pennines, but until recent times the Trent and Erewash River valleys proved to be the approximate limit of their range in central England. Changes however are taking place, perhaps because of the so-called 'greenhouse effect', which is possibly raising average temperatures sufficiently to enable southern species to breach this former barrier and expand northwards up the river valleys into central Derbyshire. Most of these newcomers (see Table 1) are as yet regarded as vagrants, but some are known to have well established colonies in counties located immediately adjacent to Derbyshire, and it is likely to be only a matter of time before they join the list of resident species. This is not a new phenomenon, for over 30 species are documented as having established themselves in the county during the present century, and most of these are now widely distributed in the lowland areas.

A broad analysis of the principal types of habitat favoured by endangered species on the Derbyshire List is as follows:-

Semi-natural deciduous woodland and scrub	38.7 %
Unimproved permanent grassland	12.9
Marshland	8.9
Alder/Willow carr	8.1
Derelict Industrial Land	8.1
Upland Moor	8.1
Lowland Heath	5.7
Parks and Gardens	4.8
Conifer Plantations	3.2
Lichens and Mosses	1.6

The importance of the remaining areas of ancient semi-natural woodland, so far as endangered species are concerned, is readily apparent, and therefore the protection and effective management of these woods are of paramount importance. Even the best of these areas are, however, only isolated fragments of what was, until the Middle Ages, a thickly wooded county, and no matter how effectively managed, the populations of endangered species dependent upon them will reach a point where their density is governed by the area of available habitat. Concentrations of colonies into one or two limited areas will always render them vulnerable and it is important to create 'wildlife corridors' linking sites of similar habitat in order to reduce the risk of destruction.

Heathland on the upland moors has undergone a considerable reduction during the present century and that remaining is faced with a variety of threats which are likely to fragment and reduce this type of habitat still further. A major additional threat is posed by global warming and reduced rainfall, which will enable greater agricultural use of these upland areas, and the moorland vegetation's having to face increasing competition from plants which at present can only thrive at lower altitudes. Climatically induced changes will exert strong pressures upon the insects of these moorland regions as well as the plants, forcing a northward movement of those species whose colonies are at present managing to survive at the southern extremity of their range.

If we examine the range of habitats favoured by species that are considered to be extinct in Derbyshire (see Table 2), it can be seen that once again it is amongst those species associated with woodland that the greatest percentage of losses lies:

Semi-natural deciduous woodland and scrub	46.9 %
Marshland	18.9
Unimproved permanent grassland	15.6
Parks and gardens	6.2
Lichens and mosses	6.2
Alder/Willow carr	3.1
Lowland heath	3.1

Wetland species of marshland, alder and willow carr also figure prominently in the list of extinctions which is not surprising considering the extent of land drainage that has taken place since the sixteenth century and continues up to the present time. Improvements to land drainage and lowering of the water table as a result of increasing demands for ground water extraction, plus the possibility of reduced rainfall, are major threats to the remaining fragments of Derbyshire's wetlands whose flora and fauna is already considerably impoverished.

All types of grassland are under threat — those in the dales of the Carboniferous limestone area, largely due to lack of management which has enabled scrub to encroach upon previously open areas, and those in the lowlands which have largely been ploughed out and converted to arable or short-term leys. Much of the remaining floristically rich grassland in the lowland areas has become established upon derelict industrial land, but this type of habitat has been largely reclaimed, and most of that which still exists is unlikely to remain undisturbed for any length of time.

Very little heathland has survived in lowland areas; most of this type of habitat survived on the ancient commons and was destroyed during the agricultural revolution of the late eighteenth/early nineteenth centuries, so those species which favoured this type of site are few and maintain a tenuous existence in the small fragmented areas which still remain.

Similarly, over a century of air pollution has decimated the county's lichen and moss communities, resulting in the loss of most of the lepidoptera whose larvae feed upon these plants. Unless the ever-increasing levels of pollutants in the atmosphere can be curbed, then it is highly likely that those species dependent upon these plants which still survive will be unable to do so for much longer.

Yellow-legged Clearwing Moth

TABLE 1. Species of Macro-lepidoptera (larger moths) regarded as vagrants.

Clay Triple-lines	*Cyclophora linearia*	Local
Tawny Wave	*Scopula rubiginata*	RDB3
Lesser Cream wave	*Scopula immutata*	Local
Treble Brown spot	*Idaea trigeminata*	Local
Balsam Carpet	*Xanthorhoe biriviata*	Na
Large Twin-spot Carpet	*Xanthorhoe quadrifasiata*	Local
Ruddy Carpet	*Catarhoe rubidata*	Nb
Devon Carpet	*Lampropteryx otregiata*	Nb
Barred Rivulet	*Perizoma bifaciata*	Local
Pinion-spotted Pug	*Eupithecia insignata*	Nb
Barred Umber	*Plagodis pulveraria*	Local
Orange Moth	*Angerona prunaria*	Local
Pale Oak Beauty	*Serraca punctinalis*	
Grass Wave	*Perconia strigillaria*	Nb
Chocolate-tip	*Clostera curtula*	Local
Black Arches	*Lymantria monacha*	Local
Clouded Buff	*Diacrisia sannio*	Local
Square-spot Dart	*Euxoa obelisca grisea*	Nb
White-line Dart	*Euxoa tritici*	
Stout Dart	*Spaelotis ravida*	Local
Square-spotted Clay	*Xestia rhomboidaria*	Nb
Great Brocade	*Eurois occulta*	Na
Bordered Gothic	*Heliophobus reticulata marginosa*	Nb
Beautiful Brocade	*Lacanobia contigua*	Local
Dog's Tooth	*Lacanobia suasa*	Local
Pod Lover	*Hadena perplexa capsophila*	Local
Varied Coronet	*Hadena compta*	

Northern Drab	*Orthosia opima*	Local
Deep-brown Dart	*Aporophyla lutulenta*	
Pale Pinion	*Lithophane hepatica*	Local
Large Ranunculus	*Polymixis flavicincta*	Local
The Sycamore	*Acronicta aceris*	Local
Reed Dagger	*Simyra albovenosa*	Nb
The Coronet	*Craniophora ligustri*	Local
Lunar-spotted Pinion	*Cosmia pyralina*	Local
Large Nutmeg	*Apamea anceps*	Local
Saltern Ear	*Amphipoea fucosa*	Local
Fen Wainscot	*Arenostola phragmitidis*	Local
The Anomalous	*Stilbia anomala*	Local
Cream-bordered Green Pea	*Earias clorana*	Nb
Burnet Companion	*Euclidia glyphica*	
Beautiful Hook-tip	*Laspeyria flexula*	Local
Pinion-streaked Snout	*Schrankia costaestrigalis*	Local

TABLE 2. Species of Macro-lepidoptera (larger moths) considered to be extinct in Derbyshire

Goat Moth	*Cossus cossus*
Hornet Clearwing	*Sesia apiformis*
White-barred Clearwing	*Synanthedon spheciformis*
Small Eggar	*Eriogaster lanestris*
Poplar Lutestring	*Tethea or*
Frosted Green	*Polyploca ridens*
The Mocha	*Cyclophora annulata*
Silky Wave	*Idaea dilutaria*
Oblique Carpet	*Orthonama vittata*
Yellow-ringed Carpet	*Entephria flavicinctata*
Lead-coloured Pug	*Eupithecia plumbeolata*
Campanula Pug	*Eupithecia denotata*
Dentated Pug	*Anticollix sparsata*
Manchester Treble-bar	*Carsia sororiata*
Waved Carpet	*Hydrelia sylvata*
Little Thorn	*Cepphis advenaria*
Small Brindled Beauty	*Apocheima hispidaria*
Satin Beauty	*Deileptenia ribeata*
Brussel's Lace	*Cleorodes lichenaria*
Privet Hawk-moth	*Sphinx ligustri*
Brown-tail	*Euproctis chrysorrhoea*
Rosy Marsh Moth	*Eugraphe subrosea*
Grey Shoulder-knot	*Lithophane ornitopus*
Sword-grass	*Xylena exsoleta*
Scarce Merveille du Jour	*Moma alpium*
Reddish Light Arches	*Apamea sublustris*
Union Rustic	*Apamea pabulatricula*
Small Rufous	*Coenobia rufa*

Waved Black	*Parascotia fuliginaria*
Marsh Oblique-barred	*Hypenodes turfosalis*
Common Fanfoot	*Herminia strigilata*

Nomenclature and arrangement follows Bradley & Fletcher (1986)
National status: Hadley (1984). Shirt (1987).

Small Argent and Sable Moth

		National Status	*No. of post-1980 Sites*
The Forester	*Adscita statices*	Nb	4

Rare and locally distributed. Unimproved acidic lowland pasture. Threatened by modern agricultural grassland management.

Cistus Forester	*Adscita geryon*	Nb	8

Uncommon and locally distributed. On rock rose in grassland areas of the Carboniferous limestone dales. Threatened by overgrazing and scrub encroachment.

Lunar Hornet Moth	*Sesia bembeciformis*		8

Uncommon and locally distributed. Larvae feeds inside trunks of sallow and willow trees growing in river valleys and lowland marshland areas. Threatened by loss of willow carr.

Currant Clearwing	*Synanthedon tipuliformis*	Nb	3

Uncommon and locally distributed. On red and blackcurrant bushes in gardens. Threatened by insecticides.

Yellow-legged Clearwing	*Synanthedon vespiformis*	Nb	2

Rare and locally distributed. Larvae feed on stumps of recently felled oak. Threatened by lack of available breeding sites since cessation of coppicing and chemical treatment of stumps after felling.

		National Status	*No. of post-1980 Sites*
Red-tipped Clearwing	*Synanthedon formicaeformis*	Nb	1

Rare; only one known site. Larvae feed in the stems of sallow and osier in the southern wetland areas. Threatened by loss of willow carr.

Large Red-belted Clearwing	*Synanthedon culiciformis*	Nb	3

Uncommon and locally distributed. Larvae feed on stumps of recently felled birch. Threatened by lack of suitable breeding sites since cessation of coppice management and chemical treatment of cut stumps.

Six-belted Clearwing	*Bembecia scopigera*	Nb	2

Rare and locally distributed. Larvae feed in roots of bird's-foot-trefoil in southern lowland areas. Threatened by loss of foodplant due to unsympathetic management of grassland habitats.

The Lappet	*Gastropacha quercifolia*		1

Rare; only one known site. On hedgerows in Trent Valley. At northern limit of its range so sensitive to local climatic conditions. Threatened by loss of habitat in southern lowlands.

Oak Lutestring	*Cymatophorima diluta hartwiegi*	Local	1

Rare; only one known site. Mature oak woodland in south of the county. Threatened by its isolation and dependence upon a single site of ancient woodland.

Grass Emerald	*Pseudoterpna pruinata atropunctaria*		4

Rare and locally distributed. Amongst broom on wasteland in lowland areas. Threatened by loss of habitat due to land reclamation and development.

Blotched Emerald	*Comibaena bajularia*	Local	5

Rare and locally distributed. Mature oak in woods and parks. At northern limit of its range so sensitive to local climatic conditions. Limited by fragmentation and relative scarcity of suitable habitat.

Small Emerald	*Hemistola chrysoprasaria*	Local	2

Rare and locally distributed. Woodland on limestone where the larval foodplant, wild clematis, is well established. At northern limit of its range. Threatened by dependence upon a single site and potential loss of foodplant as a result of woodland management works.

		National Status	*No. of post-1980 Sites*
Maiden's Blush	*Cyclophora puntaria*	Local	2

Rare and locally distributed. Young birch in lowland woods and heaths. Threatened by loss of habitat due to cessation of coppicing and destruction of lowland heath.

Cream Wave	*Scopula floslactata*	Local	3

Rare and of sporadic appearance at widely distributed sites. Deciduous lowland woods. Factors limiting its numbers are unknown; possibly at the limit of its range and sensitive to local climatic conditions.

Dwarf Cream Wave	*Idaea fuscovenosa*	Local	4

Rare and locally distributed. Wasteland sites in the Trent Valley. At the northern limit of its range. Threatened by loss of habitat due to land reclamation and development.

Satin Wave	*Idaea subsericeata*		2

Uncommon and locally distributed. Grassland in the Carboniferous limestone dales. At northern limit of its range. Threatened by overgrazing and scrub encroachment.

Small Scallop	*Idaea emarginata*	Local	5

Uncommon and locally distributed. Damp woodland, marshland and wasteland areas in the Trent Valley. Threatened by drainage and loss of habitat due to land reclamation and development.

Plain Wave	*Idaea straminata*	Local	4

Rare and of sporadic appearance. Woodland and heathland. Factors threatening its survival are unknown; possibly this moth is at the limit of its range.

Red Carpet	*Xanthorhoe munitata munitata*	Local	5

Uncommon and restricted to higher altitude moorland above 450 metres. At the southern limit of its range in the Peak District. Threatened by destruction of habitat through overgrazing, fire, soil erosion, and forestry planting.

Chalk Carpet	*Scotopteryx bipunctaria cretata*	Nb	5

Uncommon and locally distributed. Open, short-turfed areas of grassland and scree on the slopes of the Carboniferous limestone dales. Threatened by loss of open grassland due to scrub encroachment.

Lead Belle	*Scotopteryx mucronata umbrifera*		4

Rare and locally distributed. Inhabits gorse thickets on moorland and wasteland. Threatened by reclamation of marginal or derelict land for agriculture and development. At northern limit of its range.

		National Status	*No. of post-1980 Sites*
Small Argent and Sable	*Epirrhoe tristata*		3

Uncommon and locally distributed. Upland moor where heath bedstraw is well established. Threatened by overgrazing, fire, soil erosion, and forestry planting.

The Mallow *Larentia clavaria* 1

Very scarce. Road verges and wasteland in lowland areas where the larval foodplant, mallow, is established. Threatened by land reclamation, development of wasteland, and intensive road verge maintenance.

Beautiful Carpet *Mesoleuca albicillata* 5

Rare and locally distributed. Ancient woodland and parkland. Factors limiting its distribution are unknown.

Dark Spinach *Pelurga comitata* 7

Rare and locally distributed. Disturbed ground and wasteland in low-lying areas where the larval foodplant, Chenopodium, is well established. Colonies are ephemeral and rely upon a dwindling supply of habitats suitable for colonisation by Chenopodium.

Striped Twin-spot Carpet *Coenotephria salicata latentaria* 1

Locally distributed; only one known site. Restricted to upland moors above an altitude of 400 metres. At southern limit of its range in central England in the Peak District. Threatened by overgrazing, fire, soil erosion, and forestry planting

Red-green Carpet *Chloroclysta siterata* 2

Rare; only two known sites. On oak and rowan in woodland. Factors limiting its distribution are unknown but its scarcity and dependence upon only two woodland sites places it under threat.

Blue-bordered Carpet *Plemyria rubiginata* 8

Uncommon and locally distributed. Alder carr in river valleys. Loss of damp woodland habitat through drainage improvement, clearance, or conversion to conifer plantation is the principal threat.

Pine Carpet *Thera firmata* 2

Rare and locally distributed. Pine plantations. Threatened by clear felling of mature plantations.

Spruce Carpet *Thera britannica* 7

Rare and locally distributed. Spruce plantations, threatened by clear felling of mature plantations.

		National Status	*No. of post-1980 Sites*
Beech-green Carpet	*Colostygia olivata*	Local	1

Rare; only one known site. Amongst woodland and scrub on the Carboniferous limestone. Factors limiting its distribution are unknown.

Ruddy Highflyer	*Hydriomena ruberata*	Local	4

Rare and locally distributed. Damp moorland areas above 400 metres where sallows are established. Threatened by overgrazing and forestry planting. At southern limit of its range in central England.

Small Waved Umber	*Horisme vitalbata*		1

Rare; only one known site. Woodland, scrub and hedgerows, on limestone where wild clematis is well established. Threatened by loss of foodplant resulting from forestry management works. At northern limit of its range.

The Fern	*Horisme tersata*		1

Rare; only one known site. Woodland and scrub on limestone where wild clematis is well established. Threatened by loss of foodplant resulting from forestry management works. At northern limit of its range.

Pretty Chalk Carpet	*Melanthia procellata*		1

Rare; only one known site. Woodland and scrub on limestone where wild clematis is well established. Threatened by loss of foodplant resulting from forestry management works. At northern limits of its range.

Argent and Sable	*Rheumaptera hastata hastata*	Nb	3

Rare and locally distributed. Open-canopied birch woodland and scrub on moor and heath. Limiting factors are unknown, but it thrives best in open woodland and will be threatened if birch becomes too dense or is shaded out by longer-lived trees.

Scarce Tissue	*Rheumaptera cervinalis*		1

Rare and locally distributed. Dependent upon cultivated Berberis species. grown in the parks and gardens of the Derwent Valley in central Derbyshire. Threatened by insecticides and by a loss of foodplant if Berberis loses its popularity as a garden shrub.

Scallop Shell	*Rheumaptera undulata*		3

Rare and locally distributed. Damp woodland and marshland in low-lying areas. Threatened by drainage improvements and loss of wetland habitat.

		National Status	*No. of post-1980 Sites*
Brown Scallop	*Philereme vetulata*	Local	1

Rare; only one known site. Woodland or scrub on the Carboniferous limestone. Threatened by loss of larval foodplant, buckthorn.

Dark Umber	*Philerene transversata*	Local	5

Uncommon and locally distributed. Woodland or scrub on the Carboniferous limestone. Threatened by loss of larval foodplant, buckthorn.

Small Autumnal	*Epirrita filigrammaria*		4

Widely distributed on upland moors. A moorland species close to the southern limit of its range in England on the Derbyshire Moors. Threatened by overgrazing, fire, soil erosion, and forestry planting.

Barred Carpet	*Perizoma taeniata*	Na	2

Uncommon and locally distributed. Damp woodland in the limestone dales. Threatened by loss of larval foodplant (mosses) due to air pollution or prolonged drought.

Grass Rivulet	*Perizoma albulata albulata*	Local	10+

Uncommon and locally distributed. Unimproved permanent pasture or hay meadows where yellow rattle is well established. Threatened by loss of foodplant due to agricultural improvement of grassland.

Maple Pug	*Eupithecia inturbata*	Local	1

Rare; only one known site. On mature field maple in woodland. Threatened by dependence upon a single colony and scarcity of mature field maple.

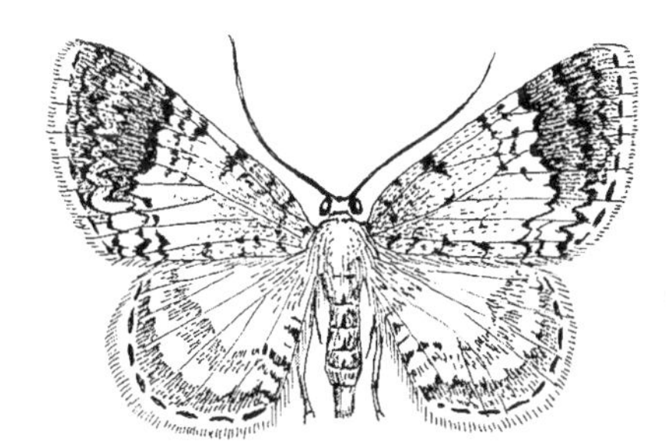

Blomer's Rivulet Moth

		National Status	*No. of post-1980 Sites*
Haworth's Pug	*Eupithecia haworthiata*	Local	1

Uncommon; only one known site. Woodland on limestone where wild clematis is well established. Threatened by its dependence upon a single colony and possible loss of foodplant resulting from woodland management works.

Valerian Pug	*Eupithecia valerianata*	Nb	2

Uncommon and locally distributed. Open areas amongst woodland or scrub in the carboniferous limestone dales. Threatened by loss of habitat due to scrub encroachment.

Netted Pug	*Eupithecia venosata venosata*	Local	5

Uncommon but widely distributed. Wasteland and ungrazed grassland where bladder campion flourishes. Threatened by loss of habitat due to land reclamation and development.

Triple-spotted Pug	*Eupithecia trisignaria*	Local	8+

Widely distributed at a low density. Wasteland, ungrazed grassland, woodland rides, etc. where wild angelica flourishes. Threatened by land reclamation, development, scrub encroachment, etc.

Freyer's Pug	*Eupithecia intricata arceuthata*		3

Rare and locally distributed. On *Cupressus* species and *Chamaecyparis* species, in parks and gardens. At northern limit of its range so probably sensitive to local climatic conditions and under threat from insecticides.

Satyr Pug	*Eupithecia satyrata*		1

Rare; only one known site. On the upland heather moors. Limiting factors are unknown but it is likely to be threatened by overgrazing, fire, soil erosion, and forestry planting.

Plain Pug	*Eupithecia simpliciata*		2

Rare and of sporadic appearance; only two known sites. Disturbed ground and wasteland areas in the lowlands where *Chenopodium* flourishes. Threatened by land reclamation, development, over-intensive road verge maintenance.

		National Status	*No. of post-1980 Sites*
Thyme Pug	*Eupithecia distinctaria constricta*	Nb	2

Uncommon and locally distributed. Short-turfed grassland where wild thyme flourishes in the limestone dales. Threatened by overgrazing and scrub encroachment.

Pimpinel Pug	*Eupithecia pimpinellata*	Nb/Local	3

Uncommon and locally distributed. Ungrazed grassland, wasteland, and disused quarries on the Carboniferous limestone. Threatened by land reclamation, development and scrub encroachment.

Oak-tree Pug	*Eupithecia dodoneata*		6

Uncommon and locally distributed. Deciduous woodland and parkland on mature oak and possibly hawthorn. Limiting factors are unknown.

Larch Pug	*Eupithecia lariciata*		4

Uncommon and locally distributed. On larch in woods and plantations. Threatened by clear felling of plantations when they reach maturity.

Dwarf Pug	*Eupithecia tantillaria*		4

Uncommon and locally distributed. On spruce in plantations. Threatened by clear felling when the trees are mature.

Sloe Pug	*Chloroclystis chloerata*		4

Uncommon but widely distributed. Hedgerows and woodland edges where blackthorn is well established. Threatened by excessive hedge trimming and removal of blackthorn during woodland management works.

Blomer's Rivulet	*Discoloxia blomeri*	Nb	2

Rare and locally distributed. Woodlands in the Carboniferous limestone dales where wych elm flourishes. Threatened by loss of larval foodplant due to Dutch Elm disease.

Dingy Shell	*Euchoeca nebulata*	Local	5

Widely distributed but uncommon. Damp woodlands in river valleys. Threatened by drainage improvements and subsequent loss of habitat.

Small White Wave	*Asthena albulata*		3

Rare and locally distributed. Deciduous woodland and scrub. Limiting factors are unknown but principal colonies are located amongst hazel scrub on woodland edges — a localised and uncommon habitat in the county.

		National Status	*No. of post-1980 Sites*
Small Yellow Wave	*Hydrelia flammeolaria*		5

Rare and locally distributed amongst alder and field maple in woodland. Limiting factors are unknown, but fragmentation of old woodland, and localised distribution of above foodplants are a contributory reason for its scarcity.

The Seraphim	*Lobophora halterata*	Local	8

Uncommon and locally distributed. A predominantly southern species inhabiting damp woodland where aspen trees are established. Factors limiting its distribution are related to the scarcity of this type of habitat, its fragmentary distribution, and possibly local climatic conditions.

Early Tooth-striped	*Trichopteryx carpinata*		1

Rare; only one known site. Damp woodland amongst alder, sallow and birch. Factors limiting distribution are unknown.

Yellow-barred Brindle	*Acasis viretata*	Local	6

Rare and locally distributed. Woodland and parkland where holly, ivy, or privet are well established. A southern species close to its northern limit in central England. Climate and isolation of suitable habitats are probable limiting factors.

Clouded Magpie	*Abraxas sylvata*	Local	5

Uncommon and locally distributed amongst elm in woodlands. Local populations have been decimated due to loss of elm trees from Dutch Elm disease.

Scorched Carpet	*Ligdia adustata*	Local	3

Rare and locally distributed. Woodland, usually on limestone, where the larval foodplant, spindle, is present. A predominantly southern species, probably sensitive to local climatic conditions, and vulnerable due to the scarcity of the foodplant.

Bordered Beauty	*Epione repandaria*		4

Rare and locally distributed. Damp woodland and willow carr in river valleys. Threatened by drainage improvement and loss of remaining fragments of wetland habitat in the Trent Valley.

Speckled Yellow	*Pseudopanthera macularia*		5

Uncommon and locally distributed. Woodland edges and scrub on the Carboniferous limestone amongst wood sage. Threatened by overgrazing and scrub encroachment.

August Thorn	*Ennomos quercinaria*	Local	3

Rare and locally distributed. Woodland and parkland in lowland areas. Limiting factors are unknown but probably at the limit of its range in central England and sensitive to local climatic conditions.

		National Status	No. of post-1980 Sites
September Thorn	*Ennomos erosaria*		8

Uncommon and locally distributed. Woodland and parkland in lowland areas. Threatened by its sensitivity to local climatic conditions and its dependence upon isolated fragments of mature deciduous woodland.

Lunar Thorn	*Selenia lunularia*	Local	7

Uncommon and locally distributed. Woodland and parkland in lowland areas. Limiting factors are probably climatic and its dependence upon isolated fragments of mature woodland and parkland.

The Annulet	*Gnophos obscurata*	Local	6

Uncommon and locally distributed. Short-turfed grassland and scree in the limestone dales. A predominantly coastal and southern insect: the Peak District represents an isolated population at its northern limit in central England. Threatened by scrub encroachment.

Grey Scalloped Bar	*Dyscia fagaria*	Local	4

Rare and locally distributed. Heathland areas on the eastern moors. Limiting factors are probably climatic due to the species being at the limit of its range. Threatened by overgrazing, fire, soil erosion, and forestry planting.

Alder Kitten	*Furcula bicuspis*	Nb	10+

Widely distributed but uncommon. Woodlands containing birch and alder in river valleys and moorland edges. Limiting factors are unknown.

Poplar Kitten	*Furcula bifida*	Local	1

Rare; only one known site. Damp woodland in the Trent Valley where poplar or aspen is well established. Threatened by drainage improvements and scarcity of mature foodplant.

Marbled Brown	*Drymonia dodonaea*		1

Rare; only one known site. Mature oak woodland in lowland areas. Threatened by dependence upon one site and scarcity of ancient oak woodland.

Lunar Marbled Brown	*Drymonia ruficornis*	Local	4

Rare and locally distributed. Mature oak woodland in lowland areas. A predominantly southern insect limited by climatic conditions and threatened by its dependence upon an uncommon and fragmented habitat.

White Satin Moth	*Leucoma salicis*	Local	10+

Widely distributed but uncommon. River valleys amongst poplar and willow carr. Threatened by drainage improvements and loss of habitat.

		National Status	*No. of post-1980 Sites*
Least Black Arches	*Nola confusalis*	Local	1

Rare; only one known site. Larvae on lichens in woods and parks. Principal threat is probably loss of lichens due to air pollution.

Light Feathered Rustic	*Agrostis cinerea*	Nb	1

Uncommon; only one known site. Rocky hillsides and short-turfed grassland in the limestone dales. An isolated population of a species at the northern limit of its range. Immediate threat is posed by scrub encroachment.

Northern Rustic	*Standfussiana lucernea*	Local	3

Uncommon but widely distributed. Upland moor and rocky hillsides in the limestone dales. At the southern limit of its range in central England. Threatened by forestry planting and scrub encroachment.

Dotted Rustic	*Rhyacia simulans*	Local	10+

Rare; and apparently declining. A general expansion in the range and population of this species occurred about 1980 and during the following years it became widely distributed in Derbyshire but a rapid decline followed and there have only been two county records since 1987. Limiting factors are unknown but climatic conditions are probably important.

Triple-spotted Clay	*Xestia ditrapezium*	Local	6

Uncommon and locally distributed. Amongst woodland and scrub in the limestone dales, river valleys and moorland edge of central Derbyshire. Limiting factors are unknown.

Heath Rustic	*Xestia agathina agathina*	Local	5

Rare and locally distributed. Upland heather moor. Threatened by overgrazing, fire, soil erosion and forestry planting.

Blue-bordered Carpet Moth

		National Status	*No. of post-1980 Sites*
Lead-coloured Drab	*Orthosia populeti*	Local	4

Rare and locally distributed. Amongst aspen in woodland. Threatened by scarcity of suitable aspen groves and isolation of existing colonies.

Striped Wainscot	*Mythimna pudorina*	Local	4

Rare and locally distributed. On common reed and canary grass in marshland areas of the Trent Valley. Threatened by fragmentation, scarcity of habitat, and improvements to land drainage. At northern limit of its range in central England.

Southern Wainscot	*Mythimna straminea*	Local	4

Rare and locally distributed. On common reed and reed canary grass in marshland areas of the Trent and Erewash Valleys. Threatened by scarcity of suitable habitats and improvements to land drainage.

The Wormwood	*Cucullia absinthii*	Nb	10

Uncommon and locally distributed. Waste land, disused railways, pit tips, etc. where the larval foodplant, wormwood, flourishes. Threatened by loss of habitat due to land reclamation and development.

The Mullein	*Cucullia verbasci*		6

Uncommon and locally distributed. Waste land, limestone grassland, parks and gardens where the larval foodplants, mullein, figwort or buddleia grow. Threatened by limited availability of wild foodplants and loss of habitat through development or scrub encroachment.

The Sprawler	*Brachionycha sphinx*		3

Rare and locally distributed. Mature oak woodland in the southern lowland areas. Limiting factors are unknown but probably sensitive to local climatic conditions and threatened by scarcity and fragmentation of available habitat.

Brindled Ochre	*Dasypolia templi*	Local	6

Uncommon and locally distributed. Disused quarries, railways, etc. on the Carboniferous limestone where the larval foodplant, hogweed, is established. Limiting factors are unknown.

Golden-rod Brindle	*Lithoma solidaginis*	Local	3

Uncommon and locally distributed. Open woodland with a ground flora of heather and bilberry on the eastern moorlands. At the southern limit of its range in central England. Threatened by overgrazing, fire and forestry planting.

		National Status	*No. of post-1980 Sites*
Tawny Pinion	*Lithophane semibrunnea*	Local	8

Uncommon and locally distributed. A recent colonist from the south whose range is at present restricted primarily to the southern lowland areas. Probably sensitive to local climatic conditions.

Red Sword-grass	*Xylena vetusta*	Local/Nb	2

Rare and locally distributed. Damp woodland and marshland. Limiting factors are unknown.

Feathered Ranunculus	*Euchmichtis lichenea lichenea*	Local	0

No post-1980 records but believed to be a rare and locally distributed resident. Rocky places and scree on the limestone dale slopes. A relict population, isolated from the predominantly coastal colonies, and threatened by loss of habitat due to scrub encroachment.

Orange Sallow	*Xanthia citrago*		6

Uncommon and locally distributed. Woods, parks and gardens where mature lime trees are established. Colonies are dependent upon isolated groups and avenues of lime trees so are threatened by the potential loss of these trees.

Barred Sallow	*Xanthia aurago*		5

Rare and locally distributed. Woodland and parkland on mature beech. The range of this predominantly southern insect has been extended northwards by the large-scale planting of beech. It however remains scarce and is probably sensitive to local climatic conditions and threatened locally by the loss of beech trees.

Dusky-lemon Sallow	*Xanthia gilvago*	Local	10

Widely distributed but uncommon. Woods, parks and hedgerows on mature elm. Threatened by loss of elm trees as a result of Dutch Elm disease.

Light Knotgrass	*Acronicta menyanthidis menyanthidis*	Local	3

Uncommon and locally distributed. Upland heather moor. A northern species at the southern limit of its range. Threatened by overgrazing, fire, soil erosion, and forestry planting.

Bird's Wing	*Dypterygia scabriuscula*	Local	4

Rare; woodland and parkland. A southern species at the northern limit of its range. Limiting factors are unknown but likely to be sensitive to local climatic conditions.

		National Status	*No. of post-1980 Sites*
The Olive	*Ipimorpha subtusa*	Local	5

Uncommon and locally distributed among aspen or poplar in woodland and parkland. Colonies are dependent upon isolated groups of Populus species, and highly vulnerable in the event of the felling of these trees.

Angle-striped Sallow	*Enargia paleacea*	Nb	3

Rare and locally distributed. Mature birch in woods and lowland heaths. A northern species close to the southern limit of its range; dependent upon large tracts of birch woodland which are often threatened by fire or felling during forestry operations.

Lesser-spotted Pinion	*Cosmia affinis*	Local	2

Rare and of sporadic appearance. On mature elm in woods, parks and hedgerows. A predominantly southern species at the limit of its range. Threatened by the loss of elm as a result of Dutch Elm disease.

White-spotted Pinion	*Cosmia diffinis*	Na/RDB3	1

Rare and of sporadic appearance. On mature elm in woods, parks and hedgerows. A southern species at the northern limit of its range in the south of the county. Threatened by loss of elm as a result of Dutch Elm disease.

The Confused	*Apamea furva britannica*	Local	3

Uncommon and locally distributed. Upland heather moor and Carboniferous limestone grassland. A northern species at the southern limit of its range. Threatened by overgrazing, fire, soil erosion and forestry planting.

Double Lobed	*Apamea ophiogranma*	Local/common	8

Uncommon and locally distributed. On reed canary grass by river banks and in marshy places. Threatened by fragmented and isolated nature of suitable habitat, land drainage, reclamation and development of wetland sites.

The Butterbur	*Hydraecia petasitis*	Local	3

Rare and locally distributed. Marshland by side of rivers and streams amongst butterbur. Threatened by land drainage, reclamation, and development of wetland sites.

Haworth's Minor	*Celaene haworthii*	Local	4

Uncommon and locally distributed. Wet areas on the upland moors amongst cotton grass. A northern species close to the southern limit of its range. Threatened by drainage, overgrazing, and forestry planting.

		National Status	*No. of post-1980 Sites*
The Crescent	*Celaena leucostigma*	Local	3

Rare and locally distributed. Damp woodland and marshland in lowland areas. Threatened by land drainage, reclamation and development of wetland sites.

Brown-veined Wainscot	*Archanara dissoluta*	Local	3

Rare and locally distributed. Amongst common reed in lowland river valleys. A southern species at the northern limit of its range. Threatened by land drainage, reclamation, and development of wetland sites.

Large Wainscot	*Rhizedra lutosa*		4

Rare and locally distributed. Amongst common reed in lowland river valleys. Threatened by land drainage, reclamation, and development of wetland sites.

Silky Wainscot	*Chilodes maritimus*	Nb/Local	3

Rare and locally distributed. Amongst reed beds in lowland river valleys. Threatened by land drainage, reclamation and development of wetland sites.

Bordered Sallow	*Pyrrhia umbra*	Local	2

Rare and locally distributed. Woodland rides and grassland on the Magnesian limestone where restharrow grows. Threatened by limited availability of suitable habitat where the larval foodplant is well established.

Scarce Silver-lines	*Bena prasinana*	Local	10+

Widely distributed but uncommon. On mature oak in woods and parks. A fairly recent colonist of the county whose population is probably unstable, sensitive to local climatic conditions, and dependent upon a limited availability of suitable habitat.

Oak Nycteoline	*Nycteola revayana*	Local	3

Rare and locally distributed. On mature oak in woods, parks, and moorland edges. Limiting factors are unknown.

Nut-tree Tussock	*Colocasia coryli*		5

Rare and locally distributed. Amongst mature beech, birch and hazel, in woods, parks and moorland edges. Limiting factors are unknown.

Scarce Silver-Y	*Syngrapha interrogationis*	Local	6

Uncommon and locally distributed. Upland heather moor. Close to the southern limit of its range on the Derbyshire moors. Threatened by overgrazing, fire, soil erosion, and forestry planting.

		National Status	*No. of post-1980 Sites*
Dark Spectacle	*Abrostola trigemina*		9

Widely distributed but uncommon. Lowland areas in the south and east of the county. Limiting factors are unknown but its distribution indicates a sensitivity to local climatic conditions.

The Blackneck	*Lygephila pastinum*	Local	2

Rare and locally distributed. On wasteland where the larval foodplant, tufted vetch, is established. A southern species at the northern limit of its range which is probably sensitive to local climatic conditions and threatened by land reclamation and development.

Small Purple-barred	*Phytometra viridaria*	Local	3

Rare and locally distributed. Short-turfed, unimproved grassland on the upland moors and Carboniferous limestone hillsides where common milkwort grows. Threatened by overgrazing and scrub encroachment.

Straw Dot	*Rivula sericealis*		3

Rare and locally distributed. Damp woodland, unimproved grassland and marshland. A southern species at the northern limit of its distribution. Sensitive to local climatic conditions and threatened by limited availability of suitable habitats.

REFERENCES

BRADLEY, J. D. & FLETCHER D. S. (1986) *An Indexed List of British Butterflies and Moths*. Kedleston Press, Orpington.

HADLEY, M. (1984) *A national review of British macro-lepidoptera*. Nature Conservancy Council, Peterborough. (Invertebrate Site Register unpublished report 46)

HARRISON, F. & STERLING M. J. (1986) *Butterflies and Moths of Derbyshire Part 2*, Derbyshire Entomological Society, Derby.

SHIRT, D. B. ed. (1987) *British Red Data Books: 2 Insects*. Nature Conservancy Council, Peterborough.

WARING, P. (1994) National moth conservation project. *News Bulletin 4*, July 1992 - December 1993. British Butterfly Conservation Society, Dunstable.

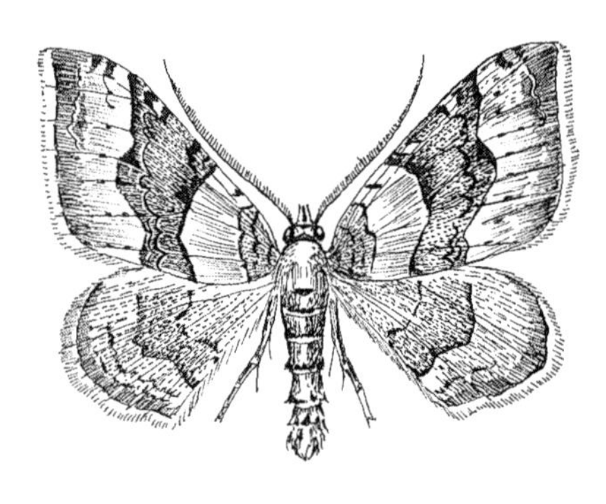

Red Carpet Moth

INDEX

English Names

Scientific Names (Genera)

THE SMALLER MOTHS (MICRO-LEPIDOPTERA)

FRED HARRISON AND KENNETH COOPER

The smaller species of moths, collectively known as the Micro-lepidoptera, are numerically the most numerous group of British Lepidoptera, and in Derbyshire so far, a total of 679 species have been identified. Their predominantly diminutive size, the close similarity between many species, and their unobtrusive life-cycles have diminished the appeal of these moths so far as most entomologists are concerned, with the result that many species are under-recorded, and much work needs to be done in order to determine their true status and distribution. An accurate assessment of their status at the present time cannot therefore be made with the same degree of confidence as applies to their larger brethren, and so the following Red Data Book list should be considered provisional. The paucity of records also limits the degree to which one can associate a species with a particular type of habitat, nor can the factors thought to be threatening its continuing existence be easily determined. As a result of this, information relating to each species is focused upon the larval foodplants, which themselves give an indication of the type of habitat in which one can expect to find the associated moth.

Nonetheless, a sufficient degree of recording has been undertaken to ascertain that some species (those readily attracted to light) are rarely observed at any of the many recording sites in which light traps are operated; certain day-flying species are scarce and rarely observed despite frequent visits to their known habitats, and that the populations of several species that are closely associated with human dwellings, warehouses, and stables, etc. have suffered considerable decline since the Victorian era.

The pattern of recording Micro-lepidoptera in Derbyshire mirrors to a much greater extent than is applicable to other families of Lepidoptera, the home neighbourhoods and favourite collecting sites of entomologists, and these areas of interest have shifted with succeeding generations of enthusiasts. During the nineteenth century, and for most of the first forty years of the twentieth century, the vast majority of records relate to the south of the county at sites in the vicinity of Willington, Repton and Burton on Trent. Dovedale was the only Peak District site regularly visited by Victorian entomologists, but during the 1920s more of the limestone dales began to receive attention and from the late 1930s onwards Lepidoptera inhabiting the northern dales and the highlands of north-west Derbyshire were studied. The 1950s saw the focus of attention switch to the north-east Derbyshire coalfield area, and later the Derwent Valley, north of Ambergate, and the eastern moorlands were frequently visited. During recent years much field work has been undertaken in the Trent Valley between Derby and Nottingham.

One species on the Derbyshire Red Data Book list of Micro-lepidoptera is of national importance and may be on the verge of extinction in the British Isles. This nationally rare species is *Euhyponoeuta stannella*, whose only known locality is situated in a very restricted area of one of the Carboniferous limestone dales. Its larval foodplant, orpine, is widely distributed in this type of location, but the moth's habitat requirements in respect of shelter, surface drainage, vegetation density and height, etc. appear to be very precise, for careful searches of the foodplant in several dales have failed to reveal the presence of colonies other than that which exists at its only known station.

The following list contains 139 species out of a Derbyshire total approaching 700 species, and it is considered that all on this list (including those for which there are no post -1980 records) are likely to still be present in the county, if searched for in suitable habitats at the appropriate time of year.

Nomenclature follows Bradley and Fletcher (1986)

The records of all those species listed in the Nature Conservancy Council Report No. 53 of July 1984: 'A Provisional National Review of the Status of British Micro-lepidoptera', by Parsons, M., have been extracted. These are described as Nationally Notable (N) or Provisional Red Data Book (PRDB) as the case may be.

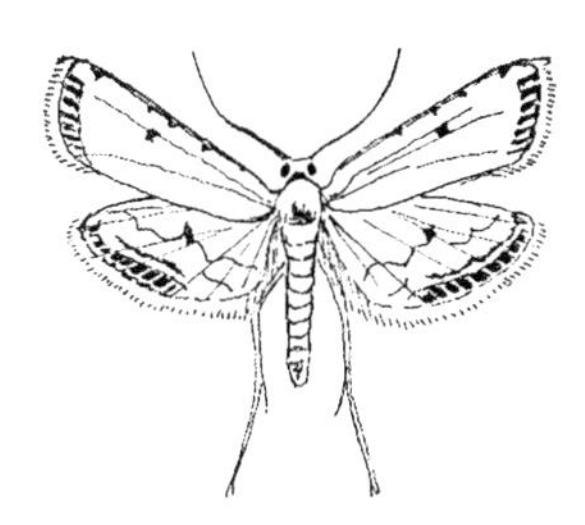

Cataclysta lemnata

	National Status	*No.of post-1980 sites*
Ectoedemia argyropeza		1

Rare; only one site. On aspen in a Carboniferous limestone dale (1989).

Stigmella speciosa		1

Rare; only one site. On sycamore in Trent Valley (1991).

Stigmella myrtillella		2

Uncommon and locally distributed. Amongst bilberry on the upland moors (1987).

Stigmella prunetorum	PRDB3	1

Rare; only one site. On blackthorn in a limestone dale (1985).

Narycia monilifera		1

Rare; only one site. Lichens growing on trees and fences in woods and old parkland (1981).

Bucculatrix nigricomella		1

Rare; only one site. On ox-eye daisy in a Carboniferous limestone dale (1988).

	National Status	*No.of post-1980 sites*
Bucculatrix frangutella		1
Rare; only one site. On buckthorn in a Carboniferous limestone dale (1986).		
Caloptilia rufipennella		1
Rare; only one site. On sycamore (1991).		
Phyllonorycter roboris		1
Rare; only one site. On oak in ancient woodland (1983).		
Phyllonorycter heegeriella		2
Rare and locally distributed. On oak in old woodlands (1985).		
Phyllonorycter junoniella		2
Locally distributed. On small plants of cowberry growing amongst heather on the upland moors (1986).		
Phyllonorycter dubitella	N	2
Rare; only two sites. On *Salix caprea* on the upland moorlands (1993).		
Phyllonorycter emberizaepennella		2
Uncommon and locally distributed. On honeysuckle in woods and parks (1991).		
Phyllonorycter leucographella		1
Rare; recent colonist in Britain. Parks and gardens on *Pyracantha* (1991).		
Phyllocnistis unipunctella		1
Rare; only one site. On poplar in woodland (1984).		
Argyresthia dilectella		1
Rare; only one site. On cultivated juniper in gardens (1984).		
Argyresthia glaucinella		1
Rare; only one site. On oak in ancient parkland in southern lowlands (1987).		
Cedestris gysseleniella		0
No post-1980 records. On pine in plantations (1977).		
Euhyponomeuta stannella	PRDB1	1
The only known site in Britain is located at a site on the county's border with Staffordshire. On orpine in a Carboniferous limestone dale (1983).		

	National Status	*No.of post-1980 sites*
Eidophasia messingiella		1

Rare and locally distributed; larvae feed on large bitter cress and hoary cress (1982).

Coleophora vitisella	N	3

Uncommon and locally distributed. On cowberry in sheltered hollows on the upland moors (1988).

Coleophora currucipennella	PRDB3	1

Rare; only one site. On oak and possibly sallow in old lowland woods (1982).

Elachista regificella	N	1

Rare; only one site. On great or hairy wood-rush in damp woodlands (1987).

Elachista kilmunella		4

Locally distributed. Possibly on *Carex* or *Eriophorum* species on the upland moors (1988).

Elachista revinctella	N	1

Rare; only one site. On *Deschampsia cespitosa* in a Carboniferous limestone dale (1985).

Semioscopis steinkellneriana		1

Rare; only one site. On blackthorn or hawthorn in the Trent Valley (1985).

Depressaria weirella		1

Rare; only one site. On cow parsley in the Trent Valley (1982).

Exaeretia allisella		5

Uncommon and locally distributed. Larvae in roots of mugwort on wasteland sites in southern and eastern Derbyshire (1984).

Agonopterix propinquella		1

Rare; only one site. On thistles in the southern lowland area (1981).

Agonopterix bipunctosa	N	1

Rare; only one site. On knapweed in a Carboniferous limestone dale (1984).

Agonopterix astrantiae	N	1

Rare; only one site. On wood sanicle in a Carboniferous limestone dale (1985).

	National Status	*No.of post-1980 sites*
Ethmia funerella	PRDB3	2

Uncommon and locally distributed. On comfrey in the limestone dales and river valleys of central Derbyshire (1993). Very local distribution nationally.

Metzneria lappella		1

Rare; only one site. On burdock growing in wasteland areas of the Trent Valley (1983).

Eulamprotes atrella		1

Rare; only one post-1980 record. On St. John's Wort in wasteland sites in the Trent Valley (1985).

Chrysoesthia drurella		1

Rare; only one site. On *Chenopodium album* on wasteland sites in the Trent Valley (1983).

Chrysoesthia sexguttella		1

Rare; only one site. On *Chenopodium album* in wasteland sites in the Trent Valley (1983).

Aristotelia ericinella		1

Rare; only one post-1980 record. Amongst heather on the upland moors (1981).

Bryotropsis affinis		1

Rare; only one-post 1980 record. On moss in the Trent Valley (1985).

Chionodes fumatella	N	1

Rare; only one site. In roots of mugwort growing on wasteland sites in the Trent Valley (1983).

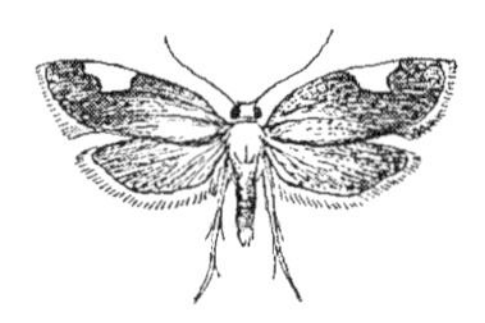

Croesia holmiana

	National Status	*No.of post-1980 sites*
Gelechia soroculella		0

No post-1980 records. On sallow at three sites (1919-75).

Scrobipalpa artemisiella	N	1

Uncommon and locally distributed. Recorded from two Carboniferous limestone dales (1955-85).

Caryocolum fraternella		1

Rare; only one record. On chickweed in the Trent Valley (1985).

Caryocolum blandella		1

Rare; only one record. On stitchwort in a Carboniferous limestone dale (1985).

Caryocolum tricolorella		2

Rare and locally distributed. On stitchwort at two widely distributed sites (1985).

Anacampsis populella		2

Uncommon and locally distributed. On aspen and poplar in ancient woodlands south of the River Trent (1985).

Brachmia blandella		1

Uncommon and locally distributed. On gorse at two sites in the Trent Valley (1976-81).

Mompha locupletella		1

Uncommon and locally distributed. On willowherb in damp places at two sites in north Derbyshire (1955-87).

Mompha raschkiella		1

Rare; only one site. On rosebay willowherb at a wasteland site in the Trent Valley (1991).

Mompha miscella		1

Uncommon and locally distributed. On rock-rose in three Carboniferous limestone dales (1928-86).

Limnaecia phragmitella		0

Rare; no post-1980 records. On reedmace at a wetland site in the Trent Valley (1972).

Pancalia leuwenhoekella	N	1

Only one site. On hairy violet in a limestone dale (1985).

	National Status	*No.of post-1980 sites*
Spuleria flavicaput		2

Widely distributed but scarce. Larvae on hawthorn. Occasional specimens recorded at sites in most areas of the county (1928-81).

Trachysmia sodaliana	N	1

Uncommon and locally distributed. In berries of purging buckthorn at a site on the Carboniferous limestone (1984).

Trachysmia maculosana		0

No post-1980 records. Uncommon and locally distributed. On bluebell in old woodland sites; central and southern Derbyshire (1963-71).

Phtheochroa rugosana		2

Uncommon and locally distributed. On white bryony in hedges, Trent Valley (1984).

Piercea alismana	N	1

Rare and locally distributed. On water plantain at a wetland site in the Trent Valley (1982).

Aethes piercei	N	1

Uncommon and locally distributed. On devil's-bit scabious in the Carboniferous limestone dales (1993).

Falseuncaria ruficiliana		1

Uncommon and locally distributed. On rock-rose in the Carboniferous limestone dales (1989).

Cochylis roseana		1

Rare; only one site. On teasel at a wasteland site in the Trent Valley (1982).

Aphelia unitana	PRDB2	2

Rare and locally distributed. On a variety of herbaceous plants such as *Angelica, Allium*, bramble, etc., in the Carboniferous limestone dales (1985).

Lozotaeniodes formosanus		2

Rare and locally distributed. A recent colonist of conifer plantations in the southern lowlands where its foodplant is Scots pine (1985).

Philedone gerningana		4

Uncommon and locally distributed. Amongst bilberry on the upland moors (1986).

	National Status	No.of post-1980 sites

Philedonides lunana 4

Uncommon and locally distributed. Amongst heather and bilberry on the upland moors (1989).

Olindia schumacherana 0

No post-1980 records. On lesser celandine and dog's mercury in the Carboniferous limestone dales (1978).

Isotrias rectifasciana 0

No post-1980 records. Foodplant unknown, possibly elm or hawthorn, at several sites in the Carboniferous limestone dales (1947-74).

Eana incanana 1

Rare and locally distributed. On bluebell in old woodlands at several widely distributed sites (1977-84).

Aleimma loeflingiana 0

No post-1980 records. On oak in old woodlands (1963-72).

Croesia holmiana 0

No post-1980 records; widely distributed but uncommon. On a variety of native deciduous trees and shrubs, mainly central and southern Derbyshire (1975-78).

Acleris comariana 1

Rare and locally distributed. On marsh cinquefoil in wetland sites 1981.

Acleris permutana PRDB3 0

No post-1980 records. On *Rosa pimpinellifolia* or blackthorn, usually in coastal locations. Several taken at Derby in 1961 were probably accidental introductions.

Acleris hastiana 0

No post-1980 records. On sallow in damp woodland, central Derbyshire (1975).

Olethreutes schulziana 1

Rare and locally distributed. On heather moorland in north-west of the county (1989).

Hedya atropunctana 1

Rare and locally distributed. On sallow and birch on the eastern moorlands (1986).

Hedya salicella 1

Uncommon and locally distributed. On sallow in damp lowland woods (1982).

	National Status	*No.of post-1980 sites*
Apotomis sororculana		1

Rare and locally distributed. On birch in old woodland site, central Derbyshire (1983).

Apotomis sauciana		5

Rare and locally distributed. Amongst bilberry on the upland moors.

Endothenia gentianeana		1

Uncommon and locally distributed. On teasel, wasteland in the Trent Valley (1983).

Endothenia marginana		1

Rare, only one site. On teasel, wasteland in the Trent Valley (1983).

Endothenia quadrimaculana		0

No post-1980 records. On marsh woundwort at a site in northern Peak District (1977).

Ancylis unguicella		1

Rare, only one site. On heather, eastern moorlands (1982).

Ancylis geminana		0

No post-1980 records. On sallow at a site in north-east Derbyshire (1977).

Ancylis laetana		0

Rare, only one site. On aspen in old woodland site, central Derbyshire (1979).

Epinotia tedella		2

Uncommon and locally distributed. On Norway spruce in conifer plantations at sites in the Peak District and eastern moorlands (1947-82).

Epinotia rubiginosana		1

Rare, only one site. On pine in a conifer plantation on the eastern moors (1986).

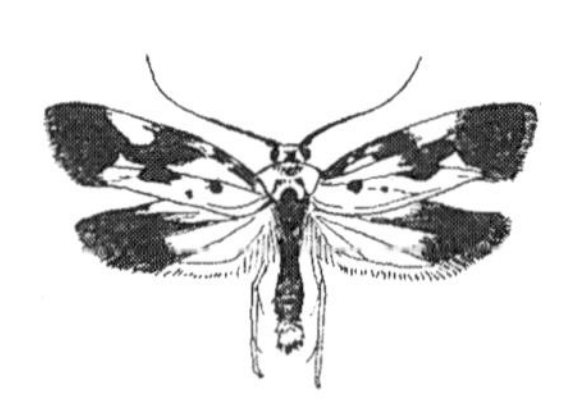

Ethmia funerella

	National Status	No.of post-1980 sites
Epinotia cruciana		0

No post-1980 records. On sallow at three sites in central and north-east Derbyshire (1963-75).

Epinotia mercuriana		1

Uncommon and locally distributed. Amongst heather and bilberry on the high moorlands of the Peak District (1987).

Epinotia caprana		1

Rare, only one site. On sallow in woodland, Trent valley (1981).

Epinotia brunnichana		1

Rare and locally distributed. On birch, hazel or sallow in woods (1983).

Rhopobota ustomaculana		0

No post-1980 records. On cowberry at several sites on the upland moors of the Peak District (1964-70).

Rhopobota stagnana		1

Uncommon and locally distributed. On field or devil's-bit scabious in the Carboniferous limestone dales (1957-85).

Rhopobota myrtillana		0

No post-1980 records. On bilberry; high moorlands of the northern Peak District (1937-56).

Zeiraphera isertana		1

Rare and locally distributed. On oak in old woodland, central and southern Derbyshire (1985).

Zeiraphera diniana		0

No post-1980 records. On larch and pine in plantations at several sites in north-east Derbyshire (1961-79).

Eucosma pupillana	N	2

Uncommon and locally distributed. In stems and roots of wormwood on wasteland sites in lowland areas of the south and east (1985).

Blastesthia posticana		1

One site on edge of eastern moorlands in Scots pine plantation (1988).

	National Status	*No.of post-1980 sites*

Rhyacionia pinicolana 1

Known from a single site in the Trent Valley (1993).

Enarmonia formosana 0

No post-1980 records. On cultivated apple and cherry trees in orchards and gardens; once common at two sites in South Derbyshire early 20th century, but only one record in last seventy years (1977).

Pammene fasciana 3

Rare and locally distributed. In acorns at three widely distributed sites (1982-86).

Cydia internana 1

Uncommon and locally distributed. On gorse at two sites in north-east Derbyshire (1962-83).

Cydia tenebrosana 1

Rare, only one recent record. Normally on rose hips but one bred from a rowan berry off eastern moorland (1984).

Cydia strobilella 1

Rare, only one site. In spruce cones from plantation on eastern moors (1981).

Cydia splendana 0

No post-1980 records. In acorns on trees at three sites in South Derbyshire (1973-76).

Dichrorampha flavidorsana 1

Uncommon and locally distributed. In roots of tansy at two wasteland sites in east Derbyshire (1963-84).

Dichrorampha consortana N 1

Rare, only one site. In stems of ox-eye daisy at a parkland site in north Derbyshire (1981).

Dichrorampha simpliciana 1

Uncommon and locally distributed. In roots of mugwort on wasteland at two sites in east Derbyshire (1962-82).

Dichrorampha gueneeana 1

Uncommon and locally distributed. In roots of yarrow and tansy on wasteland in the Trent valley (1985).

	National Status	*No.of post-1980 sites*

Dichrorampha plumbana 1

Uncommon and locally distributed. In roots of ox-eye daisy or yarrow in the Carboniferous limestone dales (1985).

Dichrorampha sedatana 1

Uncommon and locally distributed. In roots of tansy on wasteland in the Trent Valley (1984).

Alucita hexadactyla 1

Rare, only one post-1980 site. On honeysuckle in gardens mainly in the south (1990).

Agriphila latistria N 2

Uncommon and locally distributed. On grasses (*Bromus* spp.) growing on sandy soils and wasteland sites mainly in the south (1993).

Agriphila geniculea 1

Rare; only one post-1980 site. On grasses in the Carboniferous limestone dales and wasteland in the Trent valley (1956-81).

Platytes alpinella PRDB3 1

Rare; only one site. On *Tortula* spp. and other mosses. A single specimen at a site in north-east Derbyshire (1990).

Schoenobius forficella 1

Rare; only two sites. On reed sweet-grass or common reed in wetland areas of the Trent valley (1973-81).

Donacaula mucronellus 1

Rare and locally distributed. On greater pond sedge (and possibly other *Carex* spp), common reed, or reed sweet-grass in wetland areas at two sites in the Trent Valley (1973-85).

Eudonia murana 1

Rare and locally distributed. On mosses such as *Hypnum cupressiforme*, *Dicranum scoparium*, *Bryum capillare* or *Grimmia pulvinata* at four widely distributed sites in north Derbyshire (1938-81).

Nymphula stagnata 3

Uncommon and locally distributed. On bur-reed in ponds and canals at six sites, mainly in the lowland areas of Derbyshire.

	National Status	*No.of post-1980 sites*
Cataclysta lemnata		1

Uncommon and locally distributed. On duckweed in ponds and canals (1983).

Pyrausta cespitalis		0

No post-1980 records. On broad and narrow-leaved plantains at several widely distributed sites (1976-79).

Pyrausta cingulata		2

Rare and locally distributed. On thyme in the Carboniferous limestone dales (1985).

Ebulea crocealis		0

No post-1980 records. On fleabane or ploughman's spikenard in the Carboniferous limestone dales (1977).

Orthopygia glaucinalis		1

Rare and locally distributed. On dead or decaying vegetable matter in haystacks, thatch, etc., at two sites in the Trent valley (1974-85).

Pyralis farinalis		4

Rare and locally distributed. On stored cereals and refuse in barns, warehouses, etc., at four sites in the east of the county (1976-91).

Aglossa pinguinalis		1

Uncommon and locally distributed. On the refuse of cereals, hay, dung, etc., in warehouses, stables and barns at three sites (1974-85).

Achroia grisella		0

Uncommon and locally distributed. On wax in the honeycombs of beehives at two sites in the Trent Valley (1972-74).

Metriostola betulae		1

Known from a single site. On birch in southern Derbyshire (1993).

Phycita roborella		1

Rare and locally distributed. On oak or apple at three widely distributed sites (1975-81).

Dioryctria mutatella	Nb	0

No post-1980 records. On Scots pine in plantations and mixed woodland at four widely distributed sites (1969-79).

	National Status	*No.of post-1980 sites*
Hypochalcia ahenella		1

Rare and locally distributed. On rock-rose at two sites in the Carboniferous limestone dales (1926-86).

Euzophera cinerosella	Nb	2

Rare and locally distributed. On wormwood at wasteland sites in the Trent valley (1985).

Euzophera pinguis		4

Rare and locally distributed. On ash at several widely distributed sites in the south of the county (1993).

Ephestia kuehniella		0

No post-1980 records. On flour, dried herbs, etc. in flour mills and warehouses. Only two recent records from mills in east Derbyshire (1972).

Homeosoma sinuella		1

A single post-1980 record. On ribwort plantain at a site in central Derbyshire (1992).

Phycitodes maritima	N	1

Rare, only one site. On ragwort or tansy on wasteland in the Trent Valley (1982).

Platyptilia ochrodactyla	N	2

Rare and locally distributed. On tansy at wasteland sites in north-east Derbyshire and the Trent Valley (1985).

Stenoptilia saxifragae	PRDB3	3

Uncommon and locally distributed. On *Saxifraga* spp. in gardens, central and north-east Derbyshire (1994).

Pterophorus galactodactyla		0

No post-1980 record. Recorded once only from a Carboniferous limestone dale (1971).

Adana microdactyla		2

Rare and locally distributed. On hemp agrimony on the banks of rivers, ponds and canals (1983).

Leioptilus lienigianus		1

Rare, only one site. On mugwort at a wasteland site in the Trent Valley (1982).

Leioptilus tephradactyla		1

Only one post-1980 record. On golden-rod. Recorded twice this century in a

Carboniferous limestone dale (1971-92).

ACKNOWLEDGEMENTS

We are indebted to the Derbyshire Entomological Society for giving access to its data bank of records and in particular to that Society's former Recorder of Micro-lepidoptera, Mark Sterling, for his help and advice during the preparation of this section of the Red Data Book.

REFERENCES AND BIBLIOGRAPHY

BRADLEY, J. D. & FLETCHER, D. S. (1986) *An Indexed list of British Butterflies and Moths*, Kedleston Press, Orpington.

BRADLEY, J. D., TREMEWAN, W. G. & SMITH, A. (1973-79) *British Tortricoid Moths*, 2 vols, The Ray Society.

EMMET, A. M. ed. (1979) *A Field Guide to Smaller British Lepidoptera*. British Entomological & Natural History Society.

GOATER, B. (1986) *The British Pyralidae; A Guide to their Identification*. Harley Books, Colchester.

HARRISON, F. & STERLING, M. J. (1988). *The Butterflies and Moths of Derbyshire*. Pt.3.

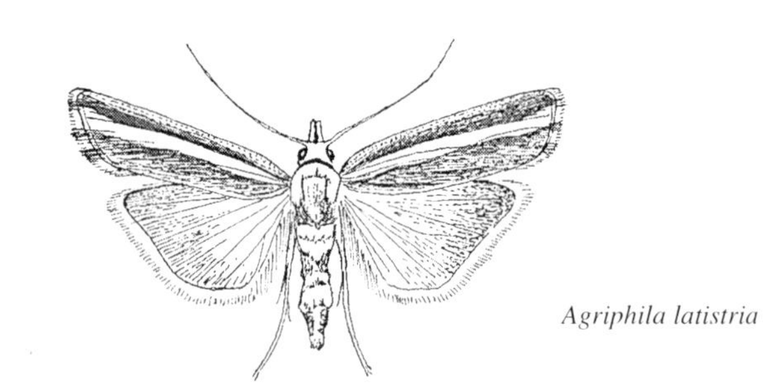

Agriphila latistria

INDEX
Scientific Names (Genera)

OTHER INSECTS

EILEEN THORPE

Insect recording in Derbyshire has been very largely the voluntary effort of amateur entomologists, each of necessity specialising in one, or only a few groups or families. Although currently more groups are being studied locally than ever before, those for which there is sufficient knowledge to attempt a chapter for this book are in the minority. The omission of many groups or families not included here should on no account be taken to imply that these contain no threatened, rare and scarce species, nor that they are insignificant. On the contrary, the national review of the aculeate hymenoptera, a group with over 500 species of ants, bees and wasps, has a higher proportion of species falling within the Red Data Book categories than any other large insect group. This is also the case for Notable species and these insects appear to have been exceptionally vulnerable to countrywide changes, with widespread local extinctions and impoverishment of faunas. Even once common bumblebees have undergone serious declines. Failure to give an account of these for Derbyshire is a serious omission. Sawflies are not included here and the very large group of parasitic hymenoptera is little known even at a national level. The flies, with around 6000 British species are also very poorly represented in this book, with an account of a single family. Other small families of flies have been studied in the county for almost as long as the hoverflies, but in general, as they are much more sparsely distributed, their records have not accumulated as quickly. It is hoped that shortly the Sorby Natural History Society will find it possible to publish a provisional account of some of these families, which we know contain species that are rare in Derbyshire. Families recorded include soldier flies, robber flies, snipe flies and snail-killing flies.

FISHES

Roy Branson

There are about 60 British species of freshwater fish, of which about half have been recorded in Derbyshire. Many of these are widespread both nationally and in the county but seven species are believed to be uncommon in Derbyshire, and one other, formerly recorded in the county, has become lost from Britain this century (Maitland,1972).

Five British fishes enjoy full legal protection. They are the sturgeon, whitefish, burbot, allis shad and vendace. Under Section 9 and Schedule 5 of the Wildlife and Countryside Act 1981 it is illegal to:

> kill, injure or capture a fish,
> possess a live or dead fish or anything derived from it,
> damage, destroy or obstruct a living site,
> disturb a fish in its living site, or
> sell a live or dead fish or anything derived from it.

The brook lamprey, salmon and spined loach also have further recognition, being listed on Annex II of the European Habitats Directive for which Special Areas of Conservation should be designated. Some fish species are legally protected by close seasons established under the Salmon and Freshwater Fisheries Act 1975 (See Parkes and Thornley,1987). Regrettably the burbot was probably lost from Britain about a decade before the act became law. The allis shad and vendace have not been recorded in Derbyshire, and the sturgeon has only been reported historically.

Although fish records are collected by angling clubs and by the National Rivers Authority they have not traditionally been the subject of interest from naturalists. Consequently it is difficult to select criteria for the species which could be considered rare or endangered and therefore suitable for inclusion in this review. Furthermore many indigenous and exotic species are widely introduced by angling organisations. The national distribution of fishes was described on a ten kilometre square basis in 1972 and has been taken as the basis for assessing the status of Derbyshire species (Maitland, 1972). Eight indigenous species were each recorded in fewer than eleven squares in Derbyshire but one of them, the rudd, has subsequently been assessed as common so only seven are described here. The salmon was not recorded in the county at the time, but has been recorded subsequently, so is also included. Three introduced species, Crucian carp *Carassius carassius* Linn., goldfish *Carassius auratus* Linn. and vendace *Coregonus vandesius* Richardson, also have fewer than eleven squares recorded but are not included here. An assessment of the status of fishes within the Trent catchment was carried out in 1991 and has been utilised to review changes since 1972 (Heaton, 1991). The national red data book for freshwater fishes had not been produced at the time of writing. Information on specific threats, population levels, future prospects etc. is scant and mentioned only where relevant. The eight indigenous species considered to be rare or endangered are described below.

The sequence and the scientific and common names of the fish described in this chapter are those used in British Freshwater Fishes (Cacutt,1979).

Brook lamprey *Lampetra planeri*

This species has been recorded in southern and western Derbyshire and the upper Derwent (Mander et al,1976), and is assessed as widely but locally distributed in the Trent catchment. It is nationally rare with only one 10 km square recorded in the county before 1960 and one after. It is thought to have been declining nationally for the last couple of decades (Cacutt,1979), and to be rare in the Midlands in 1990, but it is not considered to be under threat where it occurs.

Salmon *Salmo salar*

Salmon have been recorded in the River Trent along the county boundary between Sawley and Castle Donington during the 1980s (Templeton, 1989). County populations are assumed to be low and the species must be considered to be rare in Derbyshire. Principal threats are those associated with man, particularly pollution and artificial obstacles to migration such as weirs. As the Trent becomes cleaner and returns to a more natural temperature regime it may become suitable as a migration route for a spawning area in the Dove but populations would need to be established by introductions. This has been considered by the NRA. (Heaton, 1991).

Barbel *Barbus barbus*

This fish has been recorded in the Trent valley and lower reaches of the Dove, and has been introduced to the middle reaches of the Derwent (Templeton, pers. comm.). The barbel is locally rare with records in only one 10 km square before 1960, and four squares since, but is expanding its range.

Silver bream *Blicca bjoerkna*

This has been recorded in the Trent valley and is widespread but locally distributed in the Trent catchment. Where present, populations are strong but it is considered to be nationally rare with no county records before 1960 and records in only four 10 km squares since.

Bleak *Alburnus alburnus*

This species has been recorded in the Trent and lower Dove, and north of Chesterfield. It is locally distributed in the Trent catchment but is rare in the county with no records before 1960, and only in five 10 km squares since. However it is considered to be common where it does occur.

Spined loach *Cobitis taenia*

Local distribution is poorly known but there are records from the Trent valley. The spined loach is rare nationally and in Derbyshire with only one 10 km square record before 1960 and none since. It is very uncommon in the Trent catchment and English Nature is considering conservation action for it.

Ten-spined stickleback *Pungitius pungitius*

This species is very under-recorded but is believed to be locally rare: there are no county records before 1960 and in only five 10 km squares since. Although it is not common in the Trent catchment it is locally distributed there according to records from southern and eastern Derbyshire.

Burbot *Lota lota*

There are records in two 10 km squares in Derbyshire before 1960. These are believed to be the last burbot records from the Trent catchment in the early years of this century and the fish is believed to have become lost nationally in the early 1970s when the last specimen was recorded in Cambridgeshire (Phillips and Rix, 1985). English Nature is considering conservation action for the burbot (Heaton, 1991). This can only imply introductions from European stock.

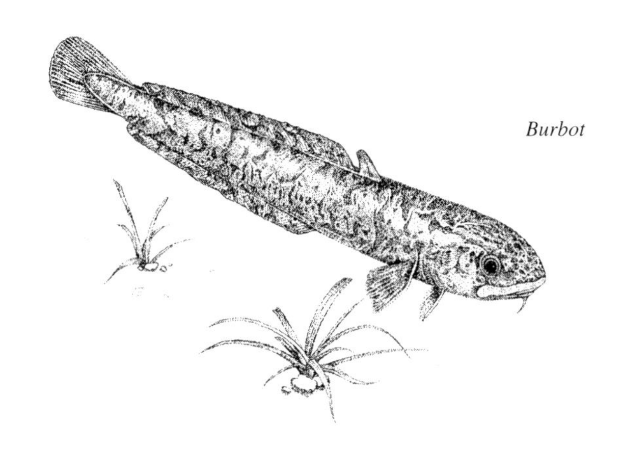

Burbot

ACKNOWLEDGEMENTS

Several people have kindly contributed time and expertise towards this paper. In particular grateful thanks are expressed to:

Andrew Heaton of the National Rivers Authority, Dr. Peter Maitland of the Fish Conservation Association, Dr. Alwyne Wheeler formerly of the British Museum (Natural History).

Grateful thanks are also expressed to Mr William M. Grange of Derby City Museum and Dr. Alan Willmot of the Derbyshire Wildlife Trust who kindly commented on the draft text.

REFERENCES

CACUTT, L. (1979) *British Freshwater Fishes: The Story of their Evolution.* Croom Helm, London.

HEATON, A. (1991) *Conservation Strategy for threatened fish in the Severn Trent region.* Internal Report, National Rivers Authority.

MAITLAND, P. S. (1972) *A Key to the Freshwater Fishes of the British Isles with Notes on their Distribution and Ecology.* Scientific Publication Number 27. Freshwater Biological Association.

MANDER, P. B., RILEY, T. H. and WHITELEY, D. (1976) *Freshwater Fishes of the Sheffield Area.* Sheffield City Museum, Sheffield.

PARKES, C. and THORNLEY, J. (1987) *Fair Game.* Pelham, London.

PHILLIPS, Roger, and RIX, Martyn. (1985) *Freshwater Fish of Britain, Ireland and Europe.* Pan, London.

TEMPLETON, R. (1989) Salmon Proves Trent Cleaner. *Stream 151.* Severn Trent Water Authority, Birmingham.

INDEX

English Names

Scientific Names (Genera)

AMPHIBIANS

Roy Branson

There are six native species of amphibians in Britain, conveniently split into two groups of three. The three newts are the great crested, the palmate and the smooth, and all occur in Derbyshire. The other group of three are frogs and toads. Derbyshire has the common toad and frog, our one omission from the group being the rare natterjack toad. There are no exotic species reported in the county.

The main threat to all of Derbyshire's amphibians is the loss of breeding sites as more rural ponds are drained or filled in for various reasons. Within any particular area a network of ponds and ditches is essential. Without a continuity of suitable habitats colonies will become isolated and increasingly vulnerable.

All of the British amphibians are protected under the Wildlife and Countryside Act 1981, some more than others. Under section 9 and Schedule 5 of the act the five Derbyshire species enjoy protection from trading: it is illegal to sell any live or dead amphibian or anything derived from one. The great crested newt has special protection: it is illegal to kill, injure or possess one and to destroy or disturb its place of shelter.

The smooth newt, common toad and frog are all threatened by habitat destruction and their legal protection is a reflection of their vulnerability. However none of them can realistically be considered to be endangered or rare in Derbyshire. The Nature Conservancy Council (now English Nature) has been monitoring the status of amphibians over the last two decades (Cooke and Scorgie,1983; Hilton-Brown and Oldham, 1991). The results are presented regionally but data have been extracted for Derbyshire and used in the following review. A county mapping scheme was set up at the Derbyshire Biological Records Centre (DBRC) at Derby Museum in the mid 1970s (Patrick, 1976) and by 1984 a general pattern was emerging (Branson and Branson, 1984). Almost 2000 records submitted to the DBRC from the mid 1970s to the end of 1990 form the base for estimating the status of amphibians in Derbyshire. They show that amphibian recorders have visited most parts of Derbyshire although some uneven coverage is apparent. Records in the centre of the county have a higher density due to the activities of the Matlock Field Club, especially Chris Monk. There are few records from the high moorland but these areas are known to be visited by naturalists so this probably means that amphibians are absent there. However a paucity of records near Chesterfield and Ashbourne probably identifies areas where no one has looked. The latest overall picture, therefore, confirms that most species are widely distributed throughout most of the county although the newts are much more sparsely distributed.

The scientific and common names are from *Reptiles and Amphibians in Britain* (Frazer,1983).

Great crested newt *Triturus cristatus*

The great crested newt has been recorded throughout Great Britain though the bulk of the records are from England (Arnold,1983). It was estimated to be sparse in Derbyshire in 1980 with a severe decrease over the previous decade. In 1990 it was widespread but not common and had declined since 1980. The DBRC records suggest a very sporadic distribution throughout the county with records in the south and east being especially isolated but this species is probably under-recorded. However in some of their established sites they can be spectacularly numerous.

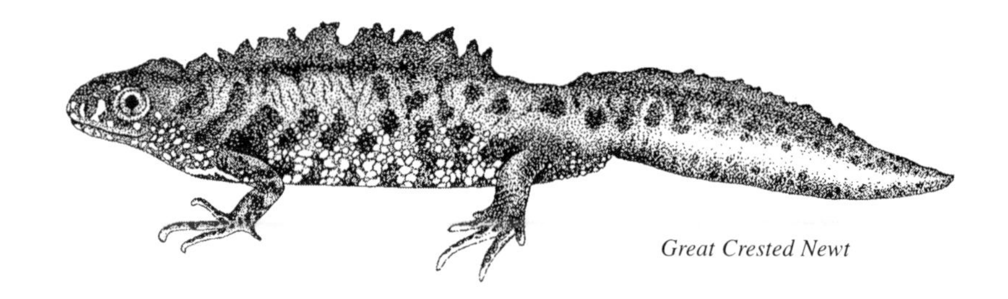

Great Crested Newt

Palmate newt *Triturus helveticus*

The palmate newt has been recorded throughout mainland Great Britain though there are fewer records from northwest, central and eastern England (Arnold, 1983). It was estimated to be sparse in Derbyshire in 1980 with no identifiable change over the previous decade. In 1990 it was also estimated to be scarce with no change during the 1980s. According to the DBRC records it is the most rare of the Derbyshire amphibians although it may be widespread in the county. Under-recorded because of the problem of identifying it. Not restricted to high ground, as many believe, but it is not common in limestone areas because it cannot tolerate high amounts of calcium carbonate in the water (Frazer, 1983).

ACKNOWLEDGEMENTS

Grateful thanks are expressed to Nicholas J Moyes, Assistant Keeper of Natural History responsible for the DBRC at Derby Museum, for his assistance with collation and interpretation of the records, and to Dr Alan Willmot of the Derbyshire Wildlife Trust for his advice on the text.

REFERENCES

ARNOLD, H. R. (1983) *Distribution Maps of the Amphibians and Reptiles of the British Isles.* Institute of Terrestrial Ecology, Huntingdon.

BRANSON, A. and BRANSON, R. (1984) The Distribution of Animals in Derbyshire. *Observations 10. Derby Natural History Society*, Derby.

COOKE, A. S. and SCORGIE, H. R. A. (1983) *The Status of the Commoner Amphibians and Reptiles in Britain.* Nature Conservancy Council, Peterborough.

FRAZER, D. (1983) *Reptiles and Amphibians in Britain.* Collins, London.

HILTON-BROWN, D. and OLDHAM, R.S. (1991) *The status of the widespread amphibians and reptiles in Britain, 1990, and changes during the 1980s.* Nature Conservancy Council, Peterborough.

PATRICK, S. (1976) Reptiles and Amphibian Distribution Mapping Scheme. *Observations 2. Derby Natural History Society*, Derby.

REPTILES

Roy Branson

There are six native British reptiles, three lizards and three snakes, and Derbyshire has two-thirds of the national species: the slow-worm, viviparous lizard, grass snake and adder. The sand lizard and the smooth snake are nationally extremely rare and confined to areas mainly in southern England. No introduced species have been recorded in Derbyshire.

Reptiles are land animals and, unlike the amphibians, can tolerate warm dry habitats. These can be extremely varied, from suburban gardens to high moorland. The biggest threats to all reptiles arise from man: either habitat destruction, disturbance or even deliberate killing.

The four Derbyshire species are protected under the Wildlife and Countryside Act 1981. Under Section 9 and Schedule 5 it is illegal to kill or injure them or to sell any live or dead reptile or anything derived from it, such as a skin.

The Nature Conservancy Council (now English Nature) has been monitoring the status of reptiles over the last two decades (Cooke and Scorgie,1983; Hilton-Brown and Oldham,1991). The results are presented regionally but data have been extracted for Derbyshire and used in the following review. Comments on British distribution are taken from Arnold (1983). A county mapping scheme for reptiles was set up at the Derbyshire Biological Records Centre (DBRC) at Derby Museum in the mid 1970s (Patrick,1976) and was reviewed in 1984 (Branson and Branson,1984). Further survey work was reported by Moyes and Branson (1990). Records submitted to the DBRC form the base for estimating the status of reptiles in Derbyshire. Fewer than 600 records have been obtained from the mid 1970s to the end of 1990, averaging only 40 sightings a year, but they show that reptile recorders have visited most parts of Derbyshire although some uneven coverage is obvious. Blank spaces on the high moorland probably mean that reptiles are absent there because the moors are known to be visited by naturalists. Other blank spaces probably identify areas where no one has looked, especially west of Derby. The latest picture suggests that although most species are widely distributed throughout the county they are very rare. Because of this, and because all species have some legal protection, all four native reptiles are included in this review.

Adder

The scientific and common names are from *Reptiles and Amphibians in Britain* (Frazer,1983).

Slow worm *Anguis fragilis*

The slow worm has been recorded throughout Great Britain including some islands, but there are more records from the south of England. It was sparsely distributed in Derbyshire in 1980, and had decreased since 1970. In 1990 it was believed to be widespread, but not common, with no change during the 1980s. It has a surprisingly widespread distribution throughout Derbyshire and may be even more widespread than the DBRC records suggest. Many of the records are concentrated around four main areas: at Little Eaton, between Ambergate and Matlock, around Lathkill Dale and around Millers Dale. Others are generally of remote single specimens though occasionally two or three may be seen together.

Viviparous lizard *Lacerta vivipara*

The viviparous lizard has been recorded widely throughout the British Isles. In 1980 it was estimated to be widespread but not common in Derbyshire, and decreasing since 1970, but its status remained unchanged during the following ten years. Records submitted to Derby museum between 1970 and 1990 are very widely distributed but those in the south and east are particularly sparse.

Grass snake *Natrix natrix*

In Britain grass snakes have only been recorded in England and Wales and records in the north of England are sparse. Indeed Derbyshire is close to the northern limit of its British range. Though its status in Derbyshire was described as widespread but not common in 1980 and 1990, it was considered to have decreased since 1970. Records at Derby Museum suggest a fairly widespread distribution in lowland Derbyshire, but the stronghold is probably between Matlock and Belper where the snakes find the Cromford Canal and adjacent grass and scrub an ideal habitat. Specimens are always elusive but away from this area they must be considered sparsely distributed.

Adder *Vipera berus*

The adder has been recorded throughout Great Britain. Its Derbyshire status was estimated to have been sparse in 1980 and decreasing over the previous decade. In 1990 it was described as widespread but not common, with no change during the 1980s. However Derbyshire records suggest that the county's principal colony is thinly distributed on moorland towards the north of the county in just three one-kilometre grid squares. Reports from the Swadlincote area date from 1977 but none have been reported for several years now so it seems likely that any colony in that area has now been lost. This much maligned snake is one of the rarest resident wild animals described in this book, with only one definite colony surviving in the face of increasing threats.

ACKNOWLEDGEMENTS

Grateful thanks are expressed to Nicholas J Moyes, Assistant Keeper of Natural History responsible for the DBRC at Derby Museum, for his assistance with the reptile data.

REFERENCES

ARNOLD, H. R. (1983) *Distribution Maps of the Amphibians and Reptiles of the British Isles*. Institute of Terrestrial Ecology, Huntingdon.

BRANSON, A. and BRANSON R. (1984) The Distribution of Animals in Derbyshire. *Observations 10. Derby Natural History Society,* Derby.

COOKE, A. S. and SCORGIE, H. R. A. (1983) T*he Status of the Commoner Amphibians and Reptiles in Britain*. Nature Conservancy Council, Peterborough.

FRAZER, D. (1983) *Reptiles and Amphibians in Britain*. Collins, London.

HILTON-BROWN, D. and OLDHAM, R. S. (1991) *The status of the widespread amphibians and reptiles in Britain, 1990, and changes during the 1980s.* Nature Conservancy Council, Peterborough.

MOYES, N. and BRANSON, R. (1990) The Derbyshire Reptile Survey: Interim Report. *Observations 16. Derby Natural History Society*, Derby.

PATRICK, S. (1976) Reptiles and Amphibian Distribution Mapping Scheme. *Observations 2. Derby Natural History Society*, Derby.

BIRDS

Roy Frost

This list includes all species which have bred at least once in Derbyshire this century and which have current populations of under fifty pairs. Introduced species are included as some, such as Greylag Goose, seem likely to be a permanent part of our avifauna.

The Derbyshire Ornithological Society (of which the writer is Rare Breeding Birds Officer) endeavours to monitor the breeding population and performance of all rare species. For some, our knowledge may be reasonably precise; for others, especially certain passerines, the situation is less clear and a degree of subjectivity is involved in the choice of species for inclusion. Lesser Spotted Woodpecker, Pied Flycatcher, Wood and Grasshopper Warblers and Siskin (all exclusions) could be near the fifty pairs base-line. Where a doubt exists, the species has been included if it is thought to be in decline.

For all species a brief history (if a recent arrival or former breeder) is given, as are some details of location within the county, habitat, populations and trends. For some of the birds of prey – an especially vulnerable group – population details are deliberately vague.

Although the number of species breeding in Derbyshire at present is probably higher than at any recent time, the rapidity with which changes in bird populations can take place leaves no room for complacency. The rapid increase among relative newcomers such as Collared Dove (present only for thirty-five years but now one of our most familiar birds) and Siskin, is countered by numerous declines of various common birds. On farmland, the most extensive habitat in the county, Rook, Turtle Dove, Little Owl, Grey Partridge, Tree Sparrow and, very markedly, Corn Bunting, all appear to have suffered as a result of agricultural changes.

Many inhabitants of broad-leaved woodlands have also been affected by changes to the woodland structure. Felling and replacement with coniferous woodland has undoubtedly led to an overall decline in numbers. However the increase in coniferous woodland has benefited several species including Firecrest, Siskin, Crossbill and, possibly, Sparrowhawk and Long-eared Owl.

Most of the gains among the breeding species, however, have concerned water and waterside birds. Although some fine wetlands have been lost (such as certain subsidence lakes in the east of the county), this has been more than balanced by the relatively recent creation of several small lakes, large reservoirs and, in particular, gravel pits. Some of the latter have been reclaimed, but it is planned that some gravel complexes will be left as water bodies with provision for nature conservation, so hopefully birds like the 'seaside quartet' (Shelduck, Ringed Plover, Oystercatcher and Common Tern), which owe their presence here almost entirely to the availability of gravel pits, will continue to breed in the county.

Human disturbance is a serious hindrance to nesting birds in some places. At our larger reservoirs the concept of multi-usage may be better on paper than in practice. Sadly the ornithological value of the new Carsington Reservoir is being limited by fishing activities in particular. Elsewhere, the recent abolition of a close fishing season at many waters is likely to threaten the breeding productivity of many wetland species. In the Peak District National Park disturbance is already very high in some places as a result of large numbers of ramblers, mountain bikers, orienteers, etc. Some of the moorland species, such as Merlin and Short-eared Owl, appear to require large areas of relatively undisturbed ground and it may be no coincidence that most of their nesting sites are in the less trodden areas. In this respect it is pleasing to note the attitude of the Peak Park Joint Planning Board in declaring a large proportion of its Eastern Moors Estate as sanctuary areas. Also encouraging is the British Mountaineering Council's decision to ask its members not to climb on certain crags in spring to avoid disturbance to Peregrines.

One recent dramatic change in breeding bird population has no direct connection with human activity. Extreme desertification of the Sahel area of Africa is believed to have reduced the numbers of several species of British summer visitors which winter or migrate through that area. The worst affected species include Sand Martin, Grasshopper Warbler, Whitethroat and Spotted Flycatcher. At the other extreme, severe winters may cause mortality among many resident birds, especially small insectivorous species such as Goldcrest and Wren and some, including Grey Heron, Barn Owl and Kingfisher, of specialised feeding habits.

Nomenclature and arrangement of species in this account are according to Voous (1977). RDB status of species is according to Batten et al. (1990).

Long-eared Owl

National Status

Mute Swan *Cygnus olor*

Although they are conspicuous residents of wetland habitats, the number of Mute Swans breeding is small. In 1987 only fourteen nesting pairs were recorded. Since that year, when the sale of lead weights for angling was prohibited, numbers have increased dramatically, with forty five pairs nesting in 1993.

National Status

Greylag Goose *Anser anser* RDB

Since 1977 a small feral breeding population in the order of five to ten pairs has become established, mainly at ornamental lakes in the south of the county.

Barnacle Goose *Branta leucopsis* RDB

A brood of young was seen at a reservoir in the north-west of the county in 1994. With an increase in the number of feral birds in the country, breeding seems likely to become regular.

Shelduck *Tadorna tadorna* RDB

Breeding first recorded in 1966 and 1981 and annually since 1984. The population of probably five to ten pairs was restricted to the Trent Valley complex, mainly at gravel pits, until 1995 when a pair nested at a lagoon in the Peak District.

Wood Duck *Aix sponsa*

A pair bred successfully in a wild state in 1984-6 at an estate in the south of the county where several had been released during the preceding years.

Mandarin Duck *Aix galericulata*

Many have been released in recent years in at least two areas in the southern half of the county. Broods of young at a small lake in 1986 and on the River Dove in 1991 and 1995 were almost certainly from these sources.

Gadwall *Anas strepera* RDB

Breeding first recorded in 1957 and annually since 1979, mainly in the Trent Valley region and in the north-east of the county; the latter is thought to be a recent westward extension of the Dukeries population. The current breeding population is probably in the order of ten to twenty pairs, scattered among a variety of wetland habitats.

Teal *Anas crecca* RDB

Recorded nesting at many localities in Derbyshire, especially in the Peak District where it is found on reservoirs, lakes and rivers. Lowland localities are fewer but more varied in habitat. In recent years up to ten broods of young have been seen in any year, a decline since the 1950s.

Garganey *Anas querquedula* RDB

Has summered in several years but breeding not proven until 1990 when a pair bred successfully in a field of rye grass in north-east Derbyshire. Nesting was considered unsuccessful in 1991-2.

Shoveler *Anas clypeata* RDB

Has bred in a variety of wetland sites in lowland Derbyshire but always rare, with never more than two pairs proven in any year. Breeding became a regular event in the late 1970s and early 1980s, but no further proof until 1992 when a pair bred on a Derbyshire Wildlife Trust reserve, with two pairs breeding elsewhere in 1993, and a successful pair on another Trust reserve in 1995.

National Status

Pochard *Aythya ferina* RDB

A regular breeding bird at a variety of water bodies every year from 1988-95 following earlier records in 1972 and 1976. Recorded from well scattered localities with a maximum of three broods in 1972 and 1994.

Red-breasted Merganser *Mergus serrator*

First bred in 1973 and in most years since 1978, with up to nine broods of young recorded in any year. Most such reports emanate from the Derwent Valley reservoirs, but has also bred in the Goyt Valley, on the Rivers Ashop and Derwent and, unexpectedly, at a Trent Valley gravel pit.

Goosander *Mergus merganser*

Although not proved to breed until 1982, has done so in most subsequent years and has increased swiftly, with a maximum of thirteen broods of young seen in 1991. All breeding records are from the Rivers Derwent, Noe, Wye and Dove. There are unconfirmed rumours of illegal persecutions.

Ruddy Duck *Oxyura jamaicensis*

Native of N. America, owing its establishment as a British breeding bird to escapes from the Wildfowl Trust in Gloucestershire. Quickly increasing after first sightings in 1963; first breeding in 1975 and annually since 1978, with probably over thirty pairs now present. Ornamental lakes and other waters with dense vegetation such as *Phragmites* and *Typha* are favoured, mainly in east, south and south-west Derbyshire. Likely to be subjected to an eradication or control programme because of hybridisation problems with the Spanish White-headed Duck.

Goshawk *Accipiter gentilis* RDB

Currently declining due to poor breeding success caused by frequent nest robberies and disturbance. Birds have also been found poisoned. Breeding was first proven in 1966 and the population, in double figures during the 1970s and 1980s, may well again be reduced to single figures. In 1989-1993 only just over two pairs a year have on average reared young, but more success has been recorded in 1994-5. Except for probable breeding records in south Derbyshire in 1983 and 1992, all sites have been woodlands in the Peak District, with a strong predilection for sites on the moorland fringe.

Buzzard *Buteo buteo*

A regular breeder in the county until the last century. An occasional and erratic breeder this century but regular since 1991, with several pairs now nesting in the Peak District and in south Derbyshire, though still subject to human disturbance.

Merlin *Falco columbarius* RDB

A moorland bird, apparently on the brink of extinction in the late 1960s and 1970s, probably due to organochlorine pesticide poisoning. Has made an excellent recovery, with a population well into double figures and generally a high fledging success rate. Now threatened by egg collectors, with many nests robbed in 1993-5.

National Status

Hobby *Falco subbuteo*

Occasional records of breeding until 1938 but then a long gap until 1975. Since 1984 breeding annually and the population has steadily increased and is now in double figures. In 1990 breeding almost certainly took place in the Peak District; otherwise all breeding sites have been in lowland farmland areas.

Peregrine *Falco peregrinus* RDB

Thought to be a regular breeder until the mid 1950s. Breeding was resumed in 1981; now with several pairs breeding at sites in both gritstone and limestone areas of the Peak District. In 1993-5 bred successfully in a box specially installed for them on a power station cooling chimney. Several recent nest robberies, believed to be mainly by pigeon fanciers and falconers.

Peregrine

Black Grouse *Tetrao tetrix* RDB

As recently as the 1960s there were several pairs in the Goyt and Derwent areas, with smaller numbers at other localities in the Peak District. However, declining in the county for years and no sightings from 1987 until May 1992 when a male and female were seen together in the High Peak, though these had possibly been released. A small and decreasing population survives on the nearby Staffordshire moors. Decline is probably mainly due to loss of habitat mosaic, coniferous afforestation and agricultural modernisation.

Quail *Coturnix coturnix* RDB

An erratic summer visitor to arable farmland areas and occasionally to other open areas such as moorlands. Breeding, most recently recorded in 1974 and 1994, is very difficult to prove. From 1980 to 1993 the number of birds recorded, mainly singing males, has varied from none to twenty nine annually, with an average of eight.

		National Status
Water Rail	*Rallus aquaticus*	

Mainly a winter visitor and passage migrant in small numbers. As a breeding bird even scarcer, although overlooked because of its secretive nature. In the last thirty years breeding has been recorded at nine sites, four of which are Derbyshire Wildlife Trust reserves. All sites are unpolluted wetlands containing dense aquatic vegetation.

Corncrake *Crex crex* RDB

Now a rare summer visitor and no longer recorded annually, while the last proven breeding was in the late 1960s. Formerly a familiar farmland bird but has declined nationally this century, probably due to farming mechanisation and improved drainage.

Oystercatcher *Haemotopus ostralegus* RDB

First nesting record in 1972 with fairly regular breeding subsequently though apparently with a low success rate. Current population between one and three pairs, all at or near gravel pits in the Trent Valley complex.

Little Ringed Plover *Charadrius dubius*

Has been colonising Britain since 1938 and first nested in Derbyshire on the banks of the River Trent in 1950, though breeding not annual until 1956. Although many breed at gravel pits a wide variety of other habitats has been used including colliery tips. Reclamation of these is probably the main reason why the population has declined from an estimated fifty-five pairs in 1979 to under fifty pairs by the early 1990s.

Ringed Plover *Charadrius hiaticula* RDB

Began breeding in Derbyshire as recently as 1978. Numbers reached a maximum of about 10 pairs in 1989 and 1993. Gravel pits are the preferred habitat but has also bred on industrial wasteland and at a lake. All sites have been in the south or east of the county.

Dunlin *Calidris alpina* RDB

Breeds on moorland from 1300 feet above sea level to the hill-tops, preferring areas dominated by cotton-grass. 1974 population estimated at ninety four pairs, but decreased considerably in recent years; current population probably about forty five pairs. Drought during the breeding season is probably the main cause of the decline.

Redshank *Tringa totanus* RDB

Current population probably twenty to forty pairs, mainly at lowland wetlands such as gravel pits, oxbows and marshes, but also found very sparingly in damp upland pasture and boggy moorland. Localised increases as a result of wetland creation have been offset by improved land drainage and conversion of pasture to arable farmland.

National Status

Common Tern *Sterna hirundo*

Breeding first proven in 1956 and annual since 1966 with a maximum of thirty-eight pairs in 1992. This is likely to decrease, at least temporarily, with the partial reclamation of two of their favoured gravel pits. Three other Trent Valley gravel pits are also used, one a Derbyshire Wildlife Trust reserve, while Common Terns now breed on rafts specially installed for them at Ogston Reservoir. Similar provision elsewhere could be beneficial.

Turtle Dove *Streptopelia turtur*

Found on lowland farmland in many parts of the county, with a stronghold in the Magnesian limestone area. Declined markedly since the early 1980s, with numbers in some areas probably reduced by 80-90 per cent. The reasons for the decline are unclear but may be connected with changed farming practices and the loss of tall hedgerows and scrub for nesting.

Barn Owl *Tyto alba* RDB

Perhaps the rarest owl in Derbyshire now, with fewer than ten pairs known, although some doubtless overlooked. Severe winters and more especially farming trends leading to large declines in vole and other rodent numbers seem the major reasons for the great decrease over the past half century or more. Open countryside with some rough grassland is the preferred habitat, while nest sites include farm buildings, holes in trees and cliff faces.

Long-eared Owl *Asio otus*

Although apparently slowly declining for a long time, recent intensive fieldwork has revealed about twenty pairs in the county, very largely in the northern half. This owl frequents woodlands of all sizes but especially coniferous plantations and dense scrub. However, unless the hunger call of the young is heard this is not an easy breeding bird to locate and the actual population may be at least twice that known.

Short-eared Owl *Asio flammeus*

Numbers appear to fluctuate considerably in accordance with the abundance of voles in its moorland breeding sites, but in a favourable year may reach double figures with the Glossop, Derwent and Goyt areas the most regularly frequented.

Nightjar *Caprimulgus europaeus* RDB

Formerly much more numerous and widespread than today. The 1977 population was considered to be under twenty pairs, mainly in the Beeley Moor – Matlock Forest area, still a regular area but with now only one to three pairs or territorial males present. Birds have also sung in north-west and north-east Derbyshire in recent years. Most recent records have been from clear-felled areas or very young plantations. Decline is national, though there are signs of a recent recovery, and likely to be mainly due to climatic changes.

Woodlark *Lullula arborea* RDB

Said to be quite common in some parts of the county in the middle of the nineteenth century but the last definite breeding recorded at Whitwell Wood in 1910.

National Status

Nightingale *Luscinia megarhynchos*

Derbyshire lies to the west of the normal breeding range, although apparently a regular visitor in small numbers about a century ago. Breeding was last proven in 1947, with only a handful of singing birds reported since, mainly in woodland and scrub on the eastern and southern fringes of the county.

Black Redstart *Phoenicurus ochruros* RDB

First bred in 1970 and has done so on several subsequent occasions, annually from 1987-90 and in 1993-5. The peak years were 1988 with four summering pairs and 1994 when six breeding pairs and another singing male were present. The most regular site is a power station and pairs have bred at another power station and three other industrial complexes, all sites being in the south and south-east of Derbyshire. Access to breeding sites is often difficult and it may well be under-recorded.

Stonechat *Saxicola torquata*

Has bred in only eight years this century, with a maximum of two successful pairs in 1981 and 1994 and four in 1995. All sites have been on gritstone moorland and rough grazing land in the Peak District.

Fieldfare *Turdus pilaris* RDB

Has bred in six years between 1969 and 1989, making Derbyshire one of the most favoured English counties. All breeding sites have been in or near the Peak Park and have included a ditch in a meadow, a limestone dale, oak woodland and a conifer plantation. One site is a Derbyshire Wildlife Trust reserve.

Firecrest *Regulus ignicapillus* RDB

A pair bred successfully in a spruce plantation in the south of the county in 1981. Singing males were found elsewhere in May 1986 and May 1989 and one was heard in suitable breeding habitat in June 1993. In 1995 breeding was proved for the second time when there were two pairs and three singing males at a locality in the north.

Red-backed shrike *Lanius collurio* RDB

This shrike is now on the verge of extinction as a British breeding bird. Last century it bred regularly in some lowland parts of the county. This century breeding records became more sporadic with the last at Cromford in 1942.

Hooded crow *Corvus corvus cornix*

A pair of this subspecies of the Carrion Crow nested on the Staffordshire side of Dovedale in 1915 and in 1959-60 one hybridised successfully with a Carrion Crow at Hardwick.

Raven *Corvus corax*

Considered a common bird some 200 years ago, but declined sharply in the nineteenth century. A pair which nested in the High Peak in 1967-8 may well have been introduced. A few presumed wanderers were seen in most subsequent years and by the mid 1990s an increasing population was present in the Peak District, with breeding recorded at several sites in both limestone and gritstone areas.

National Status

Twite *Carduelis flavirostris* RDB

Breeds on heather moorland, often feeding on adjacent farmland. In relatively small numbers for much of this century but a dramatic increase from the mid 1960s and the county population until the 1980s was probably of several hundred pairs. By 1990s much scarcer again and absent from many recent breeding sites, especially in the east of its range.

Crossbill *Loxia curvirostra*

Probably now breeds annually, usually in very small numbers; most numerous in years following large irruptions in the previous summer and autumn, with the Upper Derwent Valley and Chatsworth the most regularly recorded sites. The best ever year was 1981 when there may have been in excess of fifty pairs.

Hawfinch *Coccothraustes coccothraustes*

An erratic breeding species with several pairs breeding in a particular locality for a few years, only to then disappear for no apparent reason. Ornamental, deciduous or mixed woodland is used, both in lowland areas and in the Peak District. Recent records suggest an overall decline, though nowhere so dramatically as at Scarcliffe where a reforestation programme from the early 1960s destroyed a population estimated at fifty pairs, probably one of the densest in the country.

ACKNOWLEDGEMENTS

I am grateful to Andrew Hattersley, Richard James, Rodney Key and Derek Yalden for their comments on this article.

REFERENCES

BATTEN, L. A., BIBBY C. J., CLEMENT, P., ELLIOTT, G. D., and PORTER, R. F. eds. (1990) *Red Data Birds in Britain.* T. & A. D. Poyser, London.

BROWN, A. F. & SHEPHERD, K. B. (1991) *Breeding Birds of the South Pennine Moors.* Joint Nature Conservation Committee Report Number 7.

Derbyshire Ornithological Society (1954-94) *The Derbyshire Bird Report.*

FROST, R. A. (1978) *Birds of Derbyshire.* Moorland Publishing Co., Hartington.

GIBBONS, D. W., REID, J. B. and CHAPMAN, R. A. (1993) *The New Atlas of Breeding Birds in Britain & Ireland: 1988-1991.* T. & A. D.Poyser, London.

MARCHANT, J. H., HUDSON, R., CARTER, S. P., and WHITTINGTON, P. (1990) *Population Trends in British Breeding Birds.* British Trust for Ornithology, Tring.

VOOUS, K. H. (1977) *List of Recent Holarctic Bird Species.* Ibis supplement. London.

INDEX

English Names

Scientific Names (Genera)

MAMMALS

ROY BRANSON

There are now about 70 inland species of mammal in the British Isles of which 39 have been recorded in Derbyshire. Although concentrating on 14 rare or endangered native species, this review would not be complete without a brief mention of the 10 mammals which are rare or formerly recorded in the county but are not native, two further species which are native but have become lost recently and two native species whose presence cannot be proved.

Several of these animals are protected. Most of them are covered by the Wildlife and Countryside Act 1981, a complicated law with many conditions and exceptions, but the main features are summarised in the following table.

Animal	Illegal activities Section 9 and Schedule 5				S11 & Sch 6	
	Kill, injure or capture	Possess*	Damage, destroy or obstruct living site	Disturb animal in living site	Trade*	Trap, snare or poison
Badger						x
Bats (all)	x	x	x	x	x	x
Hedgehog						x
Otter	x	x	x	x	x	x
Red squirrel	x	x	x	x	x	x
Shrews (all)						x

* Dead, alive or anything derived from an animal, such as a pelt.

Some of these provisions also apply to the 'marginal' Derbyshire animals, such as the wildcat, dormouse, pine marten and polecat. Certain hunting restrictions apply to rabbits and hares under the Ground Game Act and to deer under the Deer Acts. Badgers are also protected under the Badger Act.

The mammal recording scheme was one of the first schemes to be set up by the Derbyshire Biological Records Centre (DBRC) at Derby Museum (Branson and Branson, 1984; Deadman, 1976) so over 16,000 records spanning nearly 20 years have been available for the current review although most are casual records rather than the results of planned surveys. Several animals, particularly bats, have been reported but records have not been submitted to the DBRC so their details are not available. However many valuable records have been transcribed from the Sorby Natural History Society's records, many of which are discussed in *Mammals of the Sheffield Area* (Clinging and Whiteley, 1980) which covers parts of north and east Derbyshire. The extent of the records and the criteria for selection of 'red data' species are discussed under each order. A national red data book for mammals has been produced by The Mammal Society (Morris, 1993), who point out the difficulties of using formal rarity categories for such mobile animals whose distributions are so poorly known.

The sequence and names of the mammals are taken from *The Handbook of British Mammals*. (Corbet and Harris,1991).

Historically the two main threats have been persecution and the loss of habitat due to man's alteration of the countryside. Pollution has also been significant but only in recent decades. Habitat loss and disturbance are probably the most important future factors.

Red Squirrel

Insectivores

Although legal protection is afforded to most of the insectivores they are generally widespread and common throughout the county (though not necessarily well recorded). The water shrew is the only one considered to be rare enough to be included in this review, having about 2% of the estimated 4430 insectivore records.

Water shrew *Neomys fodiens*

This rarely seen animal is probably widespread but not numerous. Records can only be described as sporadic but it is unlikely to exist in small isolated colonies so there are probably more of them around. Status is assumed to be rare overall, though it is not uncommon in White Peak streams (Whiteley,1985).

Bats

Eighteen species have been recorded in the British Isles in recent years though three of them have to be considered as wind-blown vagrants, and only fifteen are generally reckoned to be breeding speciès. Derbyshire probably has eight species, though one of these, Leisler's bat, was added to the list as recently as 1985, and the barbastelle has recently been added to the Staffordshire list, so could be Derbyshire's ninth in the near future.

Until the early 1980s Derbyshire had only a handful of bat records but research carried out by members of local bat groups has resulted in many more records being obtained. Although that work shows that some species are widely distributed in the county, none of them are common and some of them are extremely rare. Bats are especially difficult to identify, except in the hand. There are about 500 bat records held by Derby Museum for the period 1970 to 1991. Because of their rarity, vulnerability and special legal status, all Derbyshire species are included in this review. Comments on general British distribution are taken from Stebbings and Griffith, (1986).

All bats are seriously threatened by many of man's activities, often involving loss of habitat especially woodland, and by the use of agrochemicals in the countryside and timber preservative chemicals in buildings. Man's aggressive intolerance remains a problem too, despite extensive legal protection intended to protect them from any form of disturbance.

Lesser horseshoe bat *Rhinolophus hipposideros*

Nationally this bat now has a very limited distribution, mainly concentrated in south-western Britain, but it was formerly more widespread throughout England and Wales with well known colonies in central Derbyshire. However the last county reports were from Matlock in 1863 (Brown,1863).

Whiskered bat *Myotis mystacinus*

Widely distributed in England, Wales and Ireland and is believed to occur more densely in the south. In Derbyshire summer roosts have been found in a church porch and in farm outbuildings and hibernating specimens have been seen in Peak District limestone caves. Historically considered to be a Peak District resident, though recent records are also from the south of Derbyshire, and at least one summer roost site has been found near Derby, though it generally contains only half a dozen specimens. A cave near Matlock which historically contained large numbers in winter has been found to contain only one or two specimens. The whiskered bat is a not a common Derbyshire species and records are very scarce, many, sadly, being of dead specimens.

Brandt's bat *Myotis brandtii*

This bat so closely resembles the whiskered bat that it was not accepted as a separate species until 1970 and many bat workers still find this difficult to accept. Consequently any old field records of whiskered bat could have been for either species. Brandt's bat has been recorded throughout England and Wales. The three Derbyshire records are from sites scattered widely throughout the south of the county but have been of single specimens so no information is available on populations. Very rare in Derbyshire.

Natterer's bat *Myotis nattereri*

Natterer's bat has been recorded throughout the British Isles and the six Derbyshire records have been scattered throughout the county. All of these records have been of single specimens so no local information is available on populations. Very rare in Derbyshire.

Daubenton's bat *Myotis daubentonii*

One of the methods used to estimate the identity of Daubenton's bat is to observe its hunting technique of flying at 20 or 30 centimetres above the surface of water, catching insects flying in that zone. Although this is its principal hunting habitat it is not the only one. Nursery colonies are often in hollow trees, generally not far from water, but specimens have also been found in buildings, caves and tunnels.

Daubenton's bat is distributed throughout the British Isles. There are recent Derbyshire records at about 100 locations throughout the county. It seems likely that further research will produce far more records, although regrettably very few recent reports can be accepted as true records, because firm evidence of identification can only be obtained from hand-held specimens. Daubenton's bat must be assumed to be widespread throughout the county, but restricted to areas with reasonable stretches of open water with woodland nearby. No data is available on populations although the few roosts found by members of the Derbyshire Bat Group contained only half a dozen or so specimens. The species is considered to be uncommon along Derbyshire's waterways and rare elsewhere in the county but is probably absent from the higher land of the Peak District (Branson,1989).

Noctule *Nyctalus noctula*

Noctules are mainly woodland animals whose summer roosts and hibernation sites are often located in hollow trees although buildings are occasionally used. They are distributed throughout England, Wales and southern Scotland and are probably widespread throughout those parts of Derbyshire with suitable habitat. Many sightings are of single specimens, or sometimes half a dozen at a feeding site, but considering that they may travel twenty kilometres to feed and that 80 or 90 might come out of one tree trunk they could soon become thinly dispersed over a wide area. There are only 34 firm records and the largest roost is in the south of the county where over 90 specimens have been seen emerging from a hollow tree. Although there may be other unreported sightings the noctule can be considered to be uncommon.

Leisler's bat *Nyctalus leisleri*

Until recently this species was considered to be restricted to central England, with an isolated or relict population in southern Yorkshire. Then in 1985 members of the Sheffield Bat Group obtained firm evidence of colonies living in Derbyshire and were able to add a new species to the county list (Whiteley and Clarkson, 1985). Although considered to be a woodland species, specimens have been found in buildings too (Corbet and Harris,1991), and several Derbyshire reports have been of specimens in buildings. Only two formal records have been submitted to Derby Museum and no information is available on populations so its status is still difficult to assess, but it can be considered very rare.

Pipistrelle *Pipistrellus pipistrellus*

Pipistrelles probably roost naturally in hollow trees in summer, but the loss of suitable sites has caused alternative man-made sites to be sought. Such sites are readily available in the roofs of modern houses where nursery colonies can often be found in cavity walls, between tiles and roofing felt, and behind facias. This bat is distributed throughout the British Isles. Although it is the most widespread British and Derbyshire bat it lacks many firm historical records, but there are over 270 records, mainly in the last five years. In Derbyshire the pipistrelle is probably distributed over most of the county, including busy town centres but probably excluding the moors. Typical nursery colonies contain 10 to 30 adults but roosts of 50 or 60 have been found in the county. This species is probably reasonably common, at least by bat standards, but has declined nationally in recent years (Morris, 1993).

Brown long-eared bat *Plecotus auritus*

This is a bat of lightly wooded areas (Corbet and Harris,1991) but can be found in buildings where it usually hangs free from the central ridge beam. It occurs throughout the British Isles but is a very difficult bat to find because its whispered echo-location signal cannot be picked up by electronic bat detectors. Nevertheless occasional records of roosts in old lofts, or even dead bodies found in the street, provide evidence to suggest that this species is widespread in Derbyshire. About 70 records have been submitted and the species is assumed to be scarce or rare.

Brown long-eared Bat

Rabbits and Hares

Most of the 4000 records are casual sightings of rabbits and brown hares but about 17% are from special studies of the mountain hare colonies.

Mountain hare *Lepus timidus*

Although native to the British Isles the mountain hare exists in Derbyshire only because of deliberate introductions over a hundred years ago and it is now a Peak District speciality (Arnold,1993). Several attempts were made, particularly in 1880-82 using specimens from Perthshire, but some were more successful than others. Although the species prefers moorland habitat, colonies reported at Eyam Moor, Combs Moss, Goyts Moss and Danebower earlier this century no longer exist, but the principal colony, covering Kinder Scout and the eastern moors seems to be well established despite setbacks in severe winters. In 1984 there were about 735 mountain hares in the Peak District, varying between 200 and 1000 during the year and distributed over 246 one-kilometre grid squares (Yalden,1984). Of these squares about 80% are in Derbyshire so the county population may vary from about 430 in late autumn to about 160 after the winter losses.

Rodents

The rodents selected for more detailed descriptions are those which local researchers believe to be rare or endangered. They each have less than one per cent of the total rodent records.

Red squirrel *Sciurus vulgaris*

Although formerly widespread in the British Isles the principal centres of population are now restricted to North Wales, East Anglia, Northern England, Central Scotland and Ireland (Arnold,1993). There were a few isolated records from central Derbyshire in the 1970s, and the last stronghold in the 1980s was the coniferous woodland in the upper Derwent valley, but there have been very few recent records and this animal is now on the verge of being lost from Derbyshire. Populations are extremely low; in fact if it still survives it must be the rarest mammal in the county. Ironically some of the last Derbyshire red squirrels were eaten by goshawks — probably the country's rarest raptors (Kenward and Walls,1991). Regrettably its disappearance from Derbyshire is tantalisingly undocumented though seems to have progressed from south to north through the county. There have, apparently, been red squirrel introductions into Derbyshire, at least at Chatsworth and possibly undocumented ones elsewhere. Although several reports of red squirrels seen in Derbyshire mention melanistic (black) specimens, it has been claimed that there is no black form in the British race, although on the Continent such melanics do occur (Mallinson,1978). This might suggest that some of Derbyshire's stock may have been introduced from mainland Europe.

Yellow-necked mouse *Apodemus flavicollis*

There have been no confirmed records submitted to Derby Museum in the last two decades. Rumours of the animal's presence in eastern Derbyshire are not consistent with its habitat preference. The yellow-necked mouse prefers mature deciduous woodland on drier soils (Montgomery,1978). As most eastern Derbyshire soils are shales or clays they are usually wet and heavy. This is certainly an animal whose records would benefit from some organised research and field surveys.

Harvest mouse *Micromys minutus*

Recent records suggest that this animal is widespread in low-lying areas of east and south Derbyshire but it is suspected of having only small populations. It is unlikely that the scattered records are of isolated colonies so there may be far more than are seen. Probably never been common in the county, but almost certainly under-recorded.

Dormouse *Muscardinus avellanarius*

Another animal whose status in Derbyshire is highly dubious. There have been no records submitted to Derby Museum and, although there have been occasional rumours of colonies near Swadlincote and Sudbury, it is reasonable to assume that the dormouse has become lost from Derbyshire. The demise of this intriguing little rodent is probably due to the loss of its favourite habitat — coppiced hazel woods with plentiful honeysuckle.

Coypu *Myocastor coypus*

Native to South America but was introduced to Britain about 1929 for fur farming. When the industry slumped animals were released or escaped and were subsequently recorded in eleven different counties. There were two reports from Derbyshire in 1961 and 1975 (Lever,1977) but only one record at Derby museum, from Ogston in 1964.

Carnivores

The fox, stoat and weasel are relatively well recorded and can be considered to be widespread in the county. The badger is well recorded and has been the subject of extensive research over the last decade because of its persecution. The badger and the otter are the only two indigenous species considered in detail here but four dubious or introduced species are included.

Pine marten *Martes martes*

This exciting but little known animal was once widespread throughout the country including Derbyshire but was wiped out by Victorian gamekeepers. It declined in Derbyshire after 1800, disappearing first along the Staffordshire boundary, then from central Derbyshire, and finally from the north-west. It had gone by 1880 (Langley and Yalden,1977). A local stronghold of pine martens survived in south-west Yorkshire until at least as recently as the late 1970s (Delany,1985). It is possible that this colony may have been the source of two Derbyshire reports. One animal was reported on a May night in 1973 and a good description was supplied for another in May 1979. Both of these were from the upper Derwent valley and pine martens are known to be good travellers. Alternatively the animals could have been released or escaped from a local collection.

Polecat *Mustela putorius*

These were still widespread in 1800 but becoming rare along the Leicestershire and Nottinghamshire boundaries by 1850. They were rare throughout the county by 1880, and unrecorded by 1915, further victims of the gamekeepers (Langley and Yalden,1977). Nationally, a few animals managed to survive in the remoter areas of central Wales and they eventually began to spread back into England, until eventually, in June 1993, a dead specimen was found near Church Broughton. This was the first definite Derbyshire record for 93 years (Moyes,1994). References to the use of polecats for rabbit hunting highlight one of the biggest problems associated with estimating the status of polecats that of confusion with the ferret *Mustela furo*.

Mink *Mustela vison*

All British mink are the result of escapes or releases of this American species from fur factories. The first mink farms were established in the 1920s and escapes must have occurred soon afterwards, although breeding in the wild was not confirmed until the late 1950s (Corbet and Harris,1991). Mink were reported in Derbyshire in 1969, at Hilton and Hatton (Middleton, 1969), and there have been occasional subsequent records from around the county, but their distribution in the county is still largely unknown. The few records submitted are of casual sightings but as this animal is nocturnal it is almost certainly under-recorded.

Badger *Meles meles*

Badgers are not recorded in north-west Derbyshire but are otherwise distributed throughout the county. Typical national sett density is about 50 setts in 10 square kilometres. In the west country, where badgers are much more common, population estimates vary from 5 to 30 animals in a square kilometre but Derbyshire populations are considerably smaller. By national standards the Derbyshire badgers vary from common in central Derbyshire to scarce along the Nottinghamshire boundary (Neal, 1986).

Otter *Lutra lutra*

In Great Britain the main strongholds are Scotland, Wales, the West Country and East Anglia (Macdonald,1983). It was formerly widespread in Derbyshire but was driven out of the county (and most of the rest of England) by pollution and disturbance, though the crash in national populations did not occur until about the 1950s or 1960s. Recent Derbyshire records are of single animals except for a pair seen at Osmaston Park in 1978. It is possible that some otter reports may have been misidentified mink. There are no longer any breeding populations in the county and Derbyshire otter sightings are now extremely rare.

Wildcat *Felis silvestris*

The native species had become lost from Derbyshire before 1800, and a site in Northumberland was the only remaining English location by 1850 (Langley and Yalden, 1977). There are two remarkable reports from the upper Derwent valley during the 1970s. These cannot possibly be local wild animals and might be dismissed as cases of mistaken identity except for rumours of the release of specimens brought from Scotland.

Hoofed animals

Four species of deer and the wild boar have been reported in the county, but none of them can really be considered to be native Derbyshire animals.

Red deer *Cervus elaphus*

Although the species is native to England it has been absent from Derbyshire for many centuries. About 30 reports submitted to Derby Museum are of park, domestic, escaped or feral stock.

Fallow deer *Dama dama*

Stock has been introduced from mainland Europe on a number of occasions. There are several park herds in Derbyshire and feral populations have become established near some of them providing about 100 records.

Roe deer *Capreolus capreolus*

Originally a native British deer this animal became restricted to Scotland six or seven hundred years ago but may now be extending its range southwards again into England. Introduced populations have become well established in the south of England and in East Anglia and it is possible that these might eventually spread north and west respectively. Sixteen unconfirmed recent reports have been received at Derby Museum suggesting that colonisation of the county may be taking place, and breeding has now been confirmed in north-east Derbyshire (D. Whiteley, pers comm.).

Muntjac *Muntiacus reevesi*

Native to eastern China and Formosa. It has become established in Britain, mainly since about 1950, and now widely distributed in East Anglia, the home counties, southern England, the Welsh borders and the Midlands with some isolated records in northern England (Arnold,1993). There have been several scattered records throughout Derbyshire and it seems to be becoming established in the south and north-east of the county. The similar sized Chinese water deer (or just Water deer), *Hydropotes inermis*, has been introduced to England from China and Korea and is thinly distributed in a band from Norfolk to London. It has not been recorded in Derbyshire. One specimen of the Indian muntjac *M. muntjac* was reported from Matlock in the 1948 deer census, and a young buck of the same species which visited a tennis court on Normanton Recreation Ground, Derby on 28th February 1972 was discovered dead on nearby Warwick Avenue the following day (Lever,1977). There is some doubt over the identification, however, and these reports are far more likely to have been *M. reevesi*.

Wild boar *Sus scrofa*

Although extinct in Britain for several centuries there were reintroductions in 1826-37 (Jourdain, 1905) and a further reintroduction in 1985 when two specimens were reported over a period of several months from various farmland sites south of Matlock. Their origin remains obscure. Managers of local collections denied that any were missing. After wandering around the Crich and Whatstandwell areas for several weeks they were eventually recaptured.

Marsupials

The most well-known marsupial group is the kangaroos and the only marsupial introduced into Britain and breeding is from that group.

Red-necked wallaby *Macropus rufogriseus*

Usually considered to be a Staffordshire, rather than a Derbyshire animal but there have been several reports from the latter county since five animals escaped, or were released, from Roaches House in 1939 or 1940. A colony was established near Hoo Moor, Derbyshire, though there are no recent records from that area. Several post-war Derbyshire sightings were of lone animals, particularly of specimens dying during cold weather, so it seems unlikely that they ever bred in those areas (Lever,1977). The release of further specimens from Riber in spring 1991 resulted in several reports of sightings on the western fringe of Derby that summer, though some of the press reports were clearly hoaxes.

ACKNOWLEDGEMENTS

The data in this chapter could not have been accumulated without the contributions made by hundreds of visitors to the countryside who have diligently submitted details of their observations to the mammal recording schemes. These naturalists are too numerous to acknowledge individually, but grateful thanks are offered to Valerie Clinging, Mammal Recorder for the Sorby Society and to Nicholas Moyes, Assistant Keeper of Natural History responsible for the Derbyshire Biological Records Centre at Derby Museum.

REFERENCES

ARNOLD, H. R. (1993) *Atlas of Mammals in Britain.* HMSO, London.

BRANSON, A. and BRANSON, R. (1984) The Distribution Of Animals in Derbyshire. *Observations No 10. Derby Natural History Society,* Derby.

BRANSON, R. (1989) The distribution and status of the Daubenton's bat (*Myotis daubentoni*) in Derbyshire. *Observations No 15. Derby Natural History Society*, Derby.

BROWN, E. (1863) The Fauna and Flora of the District Surrounding Tutbury and Burton on Trent. In Mosley, O. *The Natural History of Tutbury*, London.

CLINGING, V. and WHITELEY, D. (1980). Mammals of the Sheffield Area. *Sorby Record Special Series No 3.* Sorby Natural History Society, Sheffield.

CORBET, G. B. and HARRIS, S. (1991) *The Handbook of British Mammals.* Third Edition. Blackwell, Oxford.

DEADMAN, A.(1976) Mapping Scheme for Mammal Distribution. *Observations No 2. Derby Natural History Society,* Derby.

DELANY, M. J. (1985) *Yorkshire Mammals.* University of Bradford.

GLOVER, S. (1829) *History and Gazetteer of the County of Derby.*

JOURDAIN, F. C. R. (1905) Mammals. In Page, W. (ed.) *The Victoria History of the Counties of England. Derbyshire Vol 1.* Constable, London. pp.150-158.

KENWARD, R. E. and WALLS, C. A. (1991) *Upper Derwent Squirrel and Goshawk Survey.* Institute of Terrestrial Ecology.

LANGLEY, P. J. W. and YALDEN, D. W. (1977) The decline of the rarer carnivores in Great Britain during the nineteenth century. *Mammal Review* 7: 95-116.

LEVER, C. (1977) *The Naturalised Animals of the British Isles.* Hutchinson, London.

MACDONALD, S. M. (1983) The status of the otter (*Lutra lutra*) in the British Isles. *Mammal Review.* 13: 11-23.

MALLINSON, J. (1978) *The Shadow of Extinction.* Readers Union, Newton Abbot.

MIDDLETON, J. F. (1969) *Mammals of the Derby Area.* Derby Junior Naturalists, Derby.

MONTGOMERY, W. I. (1978) Studies in the distribution of *Apodemus sylvaticus* (L.) and A. *flavicollis* (Melchior) in Britain. *Mammal Review* 8: 177-184.

MORRIS, P. A. (1993) *A Red Data Book for British Mammals.* The Mammal Society, London.

MOYES, N. J. (1994) The Fall and Rise of the Polecat in Derbyshire. *Observations No 20. Derby Natural History Society*, Derby.

NEAL, E. (1986) *The Natural History of the Badger.* Croom Helm, London.

STEBBINGS, R. and GRIFFITH, F. (1986) *Distribution and status of bats in Europe.* NERC Huntingdon.

WHITELEY, D. ed. (1985) *The Natural History of the Sheffield Area and the Peak District.* Sorby Natural History Society, Sheffield.

WHITELEY, D. and CLARKSON, K. (1985) Leisler's Bats in the Sheffield Area - 1985. *Sorby Record No 23, Sorby Natural History Society*, Sheffield.

YALDEN, D. W. (1984) The status of the mountain hare *Lepus timidus* in the Peak District. *Naturalist.*

Water Shrew

INDEX

English Names

Scientific Names (Genera)